Sustaining Rural Systems

This book examines the interplay between rural places and the competing narratives of globalization and nationalism. Through case studies from Croatia, Belgium, Australia, the USA, Argentina, Bolivia, Ecuador, Mexico, Italy and Spain, this volume highlights the contemporary status of rural change through the lens of sustainability and set within current competing narratives of globalization and economic nationalism.

The multiplicity of roles that rural communities play in economic and social systems are often overlooked in conversations about globalization and economic nationalism. Yet rural communities, economies and landscapes are closely tied to global industries, migrant flows and markets, while simultaneously subject to nationalist economic policies and strategies. The chapters in this book seek to elucidate the nuanced ties between people and industries that are at once intensely local and simultaneously tied to regional and global processes. The volume challenges us to critically examine oversimplified messaging of highly complex systems and provides insights into processes of change at local scales across major global regions.

Sustaining Rural Systems will be of great interest to upper-level students, researchers, and scholars in the areas of rural sociology, human geography and development studies. The chapters in this book were originally published as a special issue of the *Geographical Review.*

Holly R. Barcus is DeWitt Wallace Professor of Geography and Director of Asians Studies at Macalester College. Her research interests reside at the intersection of migration, identity, ethnicity and rural peripheries. She is currently Vice President and Treasurer of the International Geographical Union (IGU) Executive Committee.

William G. Moseley is DeWitt Wallace Professor of Geography, and Director of the Food, Agriculture & Society Program at Macalester College. His research interests include tropical agriculture, food security, and development policy. He currently serves on the High Level Panel of Experts of the UN Committee on World Food Security.

Sustaining Rural Systems

Rural Vitality in an Era of Globalization and Economic Nationalism

Edited by
Holly R. Barcus and William G. Moseley

LONDON AND NEW YORK

First published 2023
by Routledge
4 Park Square, Milton Park, Abingdon, Oxon OX14 4RN

and by Routledge
605 Third Avenue, New York, NY 10158

Routledge is an imprint of the Taylor & Francis Group, an informa business

British Library Cataloguing in Publication Data
A catalogue record for this book is available from the British Library

ISBN13: 978-1-032-44371-3 (hbk)
ISBN13: 978-1-032-44372-0 (pbk)
ISBN13: 978-1-003-37183-0 (ebk)

DOI: 10.4324/9781003371830

Typeset in Minion Pro
by Newgen Publishing UK

Publisher's Note
The publisher accepts responsibility for any inconsistencies that may have arisen during the conversion of this book from journal articles to book chapters, namely the inclusion of journal terminology.

Disclaimer
Every effort has been made to contact copyright holders for their permission to reprint material in this book. The publishers would be grateful to hear from any copyright holder who is not here acknowledged and will undertake to rectify any errors or omissions in future editions of this book.

Contents

Citation Information

The chapters in this book were originally published in the journal *Geographical Review*, volume 112, issue 3 (2022). When citing this material, please use the original page numbering for each article, as follows:

Introduction

Sustaining Rural Systems: Rural Vitality in an Era of Globalization and Economic Nationalism
Holly R. Barcus and William G. Moseley
Geographical Review, volume 112, issue 3 (2022), pp. 323–331

Chapter 1

Croatian Rural Futures in 2030: Four Alternative Scenarios for Postsocialist Countryside in the Newest E.U. Member State
Aleksandar Lukić, Petra Radeljak Kaufmann, Luka Valožić, Ivan Zupanc, Marin Cvitanović, Dane Pejnović and Ivan Žilić
Geographical Review, volume 112, issue 3 (2022), pp. 332–352

Chapter 2

Does Local Rural Heritage Still Matter in a Global Urban World?
Serge Schmitz and Lauriano Pepe
Geographical Review, volume 112, issue 3 (2022), pp. 353–370

Chapter 3

Young People's Visions for Life in the Countryside in Latin America
A. Cristina De La Vega-Leinert, Julia Kieslinger, Marcela Jiménez-Moreno and Cornelia Steinhäuser
Geographical Review, volume 112, issue 3 (2022), pp. 371–395

Chapter 4

Second Home Property Ownership and Public-School Funding in Wisconsin's Northwoods
Ryan Douglas Weichelt and Ezra Zeitler
Geographical Review, volume 112, issue 3 (2022), pp. 396–421

Chapter 5

Re-Turning Inwards or Opening to the World? Land Use Transitions on Australia's Western Coast
Roy Jones, Tod Jones and Colin Ingram
Geographical Review, volume 112, issue 3 (2022), pp. 422–443

Chapter 6

Exploring the Sustainability of Wilderness Narratives in Europe. Reflections from Val Grande National Park (Italy)
Giacomo Zanolin and Valerià Paül
Geographical Review, volume 112, issue 3 (2022), pp. 444–465

Chapter 7

The Emerging Mountain Imaginary of The Galician Highlands: A New National Landscape in an Era of Globalization?
Valerià Paül and Juan-M. Trillo-Santamaría
Geographical Review, volume 112, issue 3 (2022), pp. 466–492

Notes on Contributors

Holly R. Barcus, Department of Geography, Macalester College, USA.

Marin Cvitanović, Department of Life & Environmental Sciences, Bournemouth University, United Kingdom of Great Britain and Northern Ireland.

A. Cristina De La Vega-Leinert, Geography & Geology Institute, University of Greifswald, Germany.

Colin Ingram, State Government of Western Australia, Perth, Australia.

Marcela Jiménez-Moreno, Institute of Ecology, National Autonomous University of Mexico, Ciudad Universitaria, Coyoacán, Mexico.

Roy Jones, School of Design and the Built Environment, Curtin University, Australia.

Tod Jones, School of Design and the Built Environment, Curtin University Bentley Campus, Australia.

Petra Radeljak Kaufmann, Department of Geography, Faculty of Science, University of Zagreb, Croatia.

Julia Kieslinger, Institute of Geography, Friedrich-Alexander University Erlangen-Nürnberg, Germany.

Aleksandar Lukić, Department of Geography, Faculty of Science, University of Zagreb, Croatia.

William G. Moseley, Department of Geography, Macalester College, USA.

Valerià Paül, Universidade de Santiago de Compostela, Geography, Praza da Universidade, Spain.

Dane Pejnović, Retired, Department of Geography, Faculty of Science, University of Zagreb, Croatia.

Lauriano Pepe, University of Liege, UR SPHERES, Institut de Géographie, Belgium.

Serge Schmitz, University of Liege, UR SPHERES Institut de Géographie, Belgium.

Cornelia Steinhäuser, Institute of Landscape Ecology, University of Münster, Germany.

Juan-M. Trillo-Santamaría, Department of Geography, University of Santiago de Compostela, Spain.

Luka Valožić, Department of Geography, Faculty of Science, University of Zagreb, Croatia.

Ryan Douglas Weichelt, Department of Geography and Anthropology, University of Wisconsin-Eau Claire, USA.

Giacomo Zanolin, Department of Education Sciences, University of Genoa, Genova, Italy.

Ezra Zeitler, Department of Geography and Anthropology, University of Wisconsin Eau Claire, USA.

Ivan Žilić, Institute of Economics Zagreb, Zagreb, Croatia.

Ivan Zupanc, Department of Geography, Faculty of Science, University of Zagreb, Croatia.

INTRODUCTION–SUSTAINING RURAL SYSTEMS: RURAL VITALITY IN AN ERA OF GLOBALIZATION AND ECONOMIC NATIONALISM

HOLLY R. BARCUS and WILLIAM G. MOSELEY

ABSTRACT. The key organizing theme for this special issue is "Sustaining Rural Systems: Rural Vitality in an Era of Globalization and Economic Nationalism." The multiplicity of roles that rural communities play in economic and social systems are often overlooked in conversations about globalization and economic nationalism. Yet rural communities, economies and landscapes are closely tied to global industries, migrant flows and markets, while simultaneously subject to nationalist economic policies and strategies. These articles are drawn from the 2019 International Geographical Union Commission on the Sustainability of Rural Systems Colloquium and examine the interplay between rural places and the competing narratives of globalization and nationalism. The studies come from Croatia, Belgium, Australia, the USA, Argentina, Bolivia, Ecuador, Mexico, Italy and Spain. In each article, the authors seek to elucidate the nuanced ties between people and industries that are at once intensely local and simultaneously tied to regional and global processes. The articles challenge us to critically examine oversimplified messaging of highly complex systems and provide insights into processes of change at local scales across major global regions, highlighting the contemporary status of rural change through the lens of sustainability and set within current competing narratives of globalization and economic nationalism.

The terms globalization, economic nationalism, and sustainability often stand in contrast to each other. While critiques of globalization have highlighted the danger of homogenization of cultures and places for several decades, contemporary shifts in popular sentiment toward economic nationalism have spawned increasing debate over the complexities of local, regional, and global economies (Myadar and Jackson 2019). Aspects of rural places and systems, such as agricultural production and community viability, are often left out of these largely urban-centric debates, neglecting the importance, complexity, and interlinkages of rural economies and peoples to national and global systems. Sustainability is also a contentious term, policy concept, and movement that frequently is framed as being in tension with globalization and nationalism.

This special issue draws together a collection of papers addressing two key themes in rural places across the globe: recognizing demographic diversity, and land use transitions. The papers challenge us to think critically about the often oversimplified messaging of highly complex systems and to seek to provide insight into processes of change at local scales across major global regions, highlighting the contemporary status of rural change through the lens of sustainability and set within current competing narratives of globalization and economic nationalism.

Rural Geographies of Globalization, Economic Nationalism, and Sustainability

Linking the terms globalization, economic nationalism, and sustainability calls for both a recognition of contradictions and synergies across these multiscalar and complex processes. Cawley notes that "Globalization is a defining feature of recent decades. Places at great distances from each other throughout the world are linked together through flows of ideas, people, goods and investment, facilitated by advances in information and communication technologies (ICTs) and in transport ... " (2013, 1). In this framing, the global flow of goods, services, and peoples extends beyond national boundaries. However, as a counterpoint to globalization, economic nationalism suggests the prioritizing of one's own national economy over that of other countries or regions. As noted by Baughn and Yaprak, "The readiness to support nationalist economic policy is a function of the perceived economic threat posed by foreign competition. Economic nationalism is linked with personal job insecurity, authoritarianism, and intolerance of ambiguity" (1996, 759). Interestingly, rural places are often positioned at the intersection of these processes, needing to sustain or create economies that fulfil nationalist economic agendas while maintaining a position within the global market. As the papers in this special issue highlight, industries based in tourism and resources development, including heritage tourism and national park development, among other industries, often struggle to find balance between these competing agendas.

Further integrated into these narratives is the desire for "sustainability" of rural communities. Writing in a 2004 history and critique of the concept of sustainable development, Robinson " ... argue[d] for an approach to sustainability that is integrative, is action-oriented, goes beyond technical fixes, incorporates a recognition of the social construction of sustainable development, and engages local communities in new ways" (Robinson 2004, 369). Robinson both highlighted the criticisms of the concept of sustainability while reinvigorating debate about how the principles of sustainability could provide a way forward for communities and policy makers. The concepts of sustainable development and sustainability, emerging with the Brundtland Report, *Our Common Future*, in 1987 (WCED 1997), have, as Robinson (2004) pointed out, received heavy criticism, yet they are still embraced by the global community. For instance, the United Nations (U.N.), established the Sustainable Development Goals in the recent *Transforming Our World: 2030 Agenda for Sustainable Development* (United Nations 2015). Further, the U.N. is joined by its United Nations' Educational, Scientific, and Cultural Organization (UNESCO), which recently announced a social geographer as the UNESCO Chair in Global Understanding for Sustainability (IGU-UGI 2018). These actions exemplify the essential role and importance that geography and rural environments (as opposed to a singular focus on urban environments) play in achieving sustainable development goals at local, regional, and global scales (Liverman 2018; Moseley 2018). Increasingly, such calls to action include frameworks for measuring change and focus attention on rural regions in Global North and Global South countries. It is critical for governments and businesses to invest in rural regions by supporting knowledge sharing of rural challenges and opportunities for sustainable development, thereby recognizing the interdependencies of local and global rural and urban places.

This special issue grew out of a 2019 colloquium organized by the International Geographical Union (IGU) Commission on the Sustainability of Rural Systems (CSRS). Papers from 46 scholars representing 16 countries contributed to the colloquium and seven papers appear here in this special issue. An additional 12 papers are in an edited volume entitled *Rural Transformations: Globalization and its Implications for Rural People, Land, and Economies* (Barcus et al. 2022). Although the colloquium was organized around four key subthemes, in this special issue we focus on two of these themes, highlighting the importance of "local" while interrogating the relationships between local social and economic vitality, and sustainability in the context of nationalist and global trends and agendas.

Recognizing Demographic Diversity

The first subtheme, "Recognizing Rural Demographic Diversity," highlights the rapidly changing demographic landscape of rural regions. The four papers within this subtheme focus on questions of changing ethnic, racial, age, and indigenous profiles for rural communities in four specific case studies centered in Croatia, Belgium, the United States, and a cross-country comparison from four Latin American countries (Argentina, Bolivia, Ecuador, and Mexico). Some rural areas, particularly those with natural amenities, struggle to manage the social and economic disparity created by second-home ownership, tourism economies, and seasonal work (Stedman 2006; Amit-Cohen 2013; Jones and Selwood 2013). In other places, long-term economic decline and out-migration creates challenges for rapidly aging populations and their increased need for health-care options, rural poverty alleviation, and provision of services, such as schools, and employment opportunities for youth and young adults (Woods 2005; Long et al. 2013). Further challenges arise when new industries relocate to rural places and bring with them new ethnically and socially diverse populations (Barcus and Simmons 2013; Maher 2013), creating demands for rural communities to provide education, housing, and health care for new populations while simultaneously opening opportunities for growth and diversity.

In each of these sets of papers we note that the processes contributing to rural change are multiscalar and multidimensional. Although we categorize each according to a general theme, we also recognize that demographic diversity and land use changes co-occur. For example, in the first paper in this section, Lukić et al. (2022) write about four alternative future scenarios for the postsocialist countryside in the newest European Union member state of Croatia. These alternatives include 'Growth without Development,' 'Rural Renaissance,' 'Road to Nowhere,' and 'Shift.' The authors provide a systematic review of the evolution of rural development perspectives and the shortcomings of a linear development trajectory, arguing that to fully understand the future trajectories of rural places, a multiplicity of factors must be considered. This includes endogenous assets and challenges, such as locally available social and natural capital, as well as exogenous factors such as " ... endogenous development, bottom-up innovation, territorially based integrated development, social capital, sustainable development, information infrastructure, consumption, small-scale niche industries, valorization of tradition, and local embeddedness" (2 after Woods 2011). Lukić et al. (2022) walk us through the complexities of rural development situations, laying the groundwork for both understanding the potential future trajectories of rural Croatian communities and framing the complexity of rural change for the papers that follow in this special issue.

The second paper in this section tackles the question of rural heritage in the countryside of Belgium. Noting that globalization forces seem to have eroded the sense of local heritage among inhabitants of the Belgian countryside, Schmitz and Pepe (2022) propose the importance of recognizing heritage as an evolving concept in which heritage landscapes, entities, and ideals are expanding beyond material and architectural elements to include natural and intangible elements of heritage as well. The authors note the importance of changing demographics and, more broadly, how globalization has changed interests within new demographic groups, thus further influencing how local heritage and sense of local places may be lost or redefined in new contexts.

Importantly, in a multination study of young people's visions for rural living in Latin America, in the third paper, De La Vega-Leinert et al. (2022) focus on the significance of incorporating youth visions in sustainable transformation of local livelihoods and lifeways. The authors argue that decisions for leaving and staying in the countryside can have long-term consequences for rural development and migratory policies, as well as for environmental sustainability practices. The studies in this paper come from Argentina, Bolivia, Ecuador, and Mexico and explore how young people perceive past, present, and future countrysides, arguing that the aim at fostering sustainability pathways needs to embrace young people's requirements, aspirations, and visions, and support them both in developing their countrysides and enabling their (im)mobilities through broader interconnected communities. Much as in Schmitz and Pepe (2022), De La Vega-Leinert et al. (2022) find that new or emerging groups within rural spaces may have significantly different visions for the future, or past, of a particular rural place.

The final paper in this theme creates a pivot point across the demographic and land use change themes. Weichelt and Zeitler (2022) provide an insightful untangling of second home ownership and public funding in rural Wisconsin, USA. They discuss the dynamic shift of funding for local communities, evolving from primary dependence on real estate taxes from year-round local residents to second home owners and land controllers. They tackle the sticky and often contentious question of land control, who owns the most valuable land, and the implications of this ownership on local taxes, and by extension, local institutions, such as schools, that are dependent on these taxes for survival. Collectively, these four papers illustrate the impacts of global processes and imaginations on local economies and demographics.

Land Use Transitions

The second theme in this special issue is "Land Use Transitions." The restructuring of global, and by extension, rural, economies, such as " … the liberalization of global trade and the increasingly 'foot-loose' nature of economic enterprises … ." (Woods 2005, 63), creates new labor demands (or deficiencies), and greater requirements for both physical and technological infrastructure, fostering different sets of land uses and land valuing. One of the primary processes of restructuring has been the shift from production-oriented rural economies to consumption-oriented rural economies (Woods 2005). Postproductive rural landscapes in the Global North herald less favorable economic prospects for production-based, or "traditional," rural economies such as mining, forestry, fishing, and agriculture. According to Woods (2005), postproductive, consumption-based rural economies are often based on tourism or other service-oriented development, such as telephone call centers. New economies and land uses are cause for a range of conflicts between different user and resident groups, as well as challenging questions of land ownership, historic land rights and voice across different resident groups (Barcus and Smith 2015). For example, conflicts over rangelands in the western United States between traditional ranchers and new residents has resulted in a range of new policy implications for these areas (see, for example, Loffler and Steinicke 2006). Globally, questions of tourism development and who benefits and loses from such development have also garnered attention (see, for example, Chio 2014). Concerns over the conservation and preservation of natural areas, particularly in areas where such preservation might be perceived to impede other types of economic development, can lead to highly charged social environments. One such example is the recent controversy over copper-nickel sulfide mining in the Boundary Waters Canoe Area in the northern United States (Forgrave 2017). The traditional economies of northern Minnesota, including mining and logging, stand in contrast to a view of nature as needing protection from development. Similar conflicts over preferred land uses arise between tourists or short-term, summer residents and long-term, full-time residents, with summer residents often favoring recreation amenities and conservation of wild areas, while full-time residents may favor resource development with greater economic opportunities, such as logging or mining.

The three papers in this section (and the last three papers in this special issue) highlight land use transitions and the reimagination of land purposes. Jones et al. (2022) focus their work on land use transitions occurring along a 100 km section of the Western Australian coast. They evaluate the historical evolution of land use from pre-European settlement to the present, highlighting ongoing tensions between local desires for stability with external pressures for change, noting that each category also includes competing or conflicting perspectives. External pressures, they describe, come in the form of pressure for economic development, environmental conservation, acknowledgment and formalization of indigenous land ownership, and management rights. The authors adeptly walk readers through the historical transitions of the coastal area, drawing connections between achieving rural sustainability and vitality at a local scale and engaging global processes.

The last two papers take aim at the evolution of wilderness narratives and narratives of national landscape within local imaginations. In Italy, Zanolin and Paül (2022) explore the evolution of national park narratives from one of conservation to a greater focus on utilitarian uses, tourism in particular, influencing the national park's policy making, specifically focusing on Val Grande National Park. The authors seek to elucidate the competing ideas of designating wilderness, or wild areas, with the development of such areas for tourism based on the designation of wilderness. They conclude that "wilderness needs to be reconceptualized so that contemporary European protection policies might become more effective, and we may use our knowledge of nature to promote sustainable development."

In the final paper of this special issue, Paül and Trillo-Santamaría (2022) investigate the emergence of the Galician highlands in Spain in the context of nationalism and the geographic narratives that are often attached to the national imaginary of mountains. The authors evaluate the changing ways in which mountains in Western cultures are envisioned, probing the construction of national landscape imaginaries. In their paper, the authors utilize Spain's Galician Highlands to explore how national landscapes are embedded in historical understandings of place and idealized visions of land and identity. They further argue that in addition to shifts from coastal farming landscapes in Galicia to an inclusion of mountains in this same vision, a secondary branding of the Galician Highlands with mountains is being utilized to promote tourism for nonlocal tourists. They conclude that " . . . the ultimate construction of a new national landscape for Galicia based on mountains, although using some national attributes . . . seems more related to the need of positioning Galicia internationally, embracing an inherent global dimension." Thus, this last paper forefronts the intersection of globalization and landscape imagery to link local imaginaries of traditional landscapes with the desire to promote landscapes perceived to appeal to global tourism.

Conclusions

Rural vitality in an era of globalization and nationalism requires continued examination of the interconnectedness and sustainability of rural places. The papers included in this special issue bring into focus the complex and multi-scalar systems that create opportunities for new imaginations of rurality, rural economies, rural infrastructure development and the shifting and sometimes contested identities of rural places. Strategies for embracing sustainability at the local level, nested within competing nationalist and globalization narratives, are described in each of these papers, exhibiting at once the unique socio-geographic context of each place while underscoring the interconnectedness and interdependence of land use, demographics, and broader national and global processes.

Collectively, the authors presented these papers in July of 2019, several months before COVID-19 emerged and imprinted itself on the evolution and development trajectories of all places across the globe, not the least rural places. In these past two years, with lockdowns, work-at-home orders, and new ways of thinking about space and place, the debates around globalization and nationalism have intensified and underscore the imperative for understanding the multiplicity of ways local communities are able to respond and adapt to such global shocks. While the effects of COVID-19 lockdowns and travel restrictions are only beginning to be studied, these measures no doubt restricted global tourism flows to rural areas while simultaneously opening opportunities for telework and the use of once seasonal homes for longer-term residences. As such, the pandemic has been another example of the dynamic interaction between globalization, economic nationalism, and sustainability that often plays out across rural landscapes.

Acknowledgments

We deeply appreciate all the contributors to this special issue, the reviewers, and Dave Kaplan, Editor, for their ideas, insights and commitment to completing the project. We also thank our funders, NSF Award #1853832, Macalester College and the University of Wisconsin Eau Claire, for supporting the initial Colloquium from which these articles emerged.

Funding

This work was supported by the Macalester College; NSF; University of Wisconsin Eau Claire.

Disclosure statement

No potential conflict of interest was reported by the author(s).

ORCID

Holly R. Barcus http://orcid.org/0000-0002-3841-5725

William G. Moseley http://orcid.org/0000-0003-0662-962X

References

Amit-Cohen, I. 2013. Heritage Landscape Fabrics in the Rural Zone: An Integrated Approach to Conservation. In *Globalization and New Challenges of Agricultural and Rural Systems: Proceedings of the 21st Colloquium of the CSRS of the IGU*, edited by D.-C. Kim, A. M. Firmino, and Y. Ichikawa, 79–88. Japan: Nagoya University Press.

Barcus, H. R., R. Jones, and S. Schmitz, editors. 2022. *Rural Transformations: Globalization and Its Implications for Rural People, Land, and Economies*. London: Taylor & Francis Group. (forthcoming).

Barcus, H. R., and L. Simmons. 2013. Ethnic Restructuring in Rural America: Migration and the Changing Faces of Rural Communities in the Great Plains. *Professional Geographer* 65 (1):130–152. doi:10.1080/00330124.2012.658713.

Barcus, H. R., and L. J. Smith. 2015. Facilitating Native Land Reacquisition in the Rural United States through Collaborative Research and Geographic Information Systems. *Special Issue of Geographical Research* 54(2):118–128. doi:10.1111/1745-5871.12167.

Baughn, C. C., and A. Yaprak. 1996. Nationalism: Conceptual and Empirical Development. *Political Psychology* 17(4):759–778. doi:10.2307/3792137.

Cawley, M. 2013. Introduction: Context and Contents. In *The Sustainability of Rural Systems: Global and Local Challenges and Opportunities*, edited by M. Cawley, A. M. D. S. M. Bicalho, and L. Laurens, 1–14. Galway, Ireland: CSRS of the IGU and the Whitaker Institute, National University of Ireland Galway.

Chio, J. 2014. *A Landscape of Travel: The Work of Tourism in Rural Ethnic China*. Seattle: University of Washington Press.

De La Vega-Leinert, A. C., J. Kieslinger, M. Jiménez-Moreno, and C. Steinhäuser. 2022. Young People's Visions for Life in the Countryside in Latin America. *Geographical Review*, 112(3), 371–395. doi:10.1080/00167428.2021.1925897.

Forgrave, R. 2017. In Northern Minnesota, Two Economies Square Off: Mining Vs. Wilderness. *New York Times*, October 12. https://www.nytimes.com/2017/10/12/magazine/in-northern-minnesota-two-economies-square-off-mining-vs-wilderness.html.

IGU-UGI. 2018. UNESCO Honours Geography and Establishes the UNESCO Chair in Global Understanding for Sustainability. International Geographical Union. https://igu-online.org/unesco-honours-geography-and-establishes-the-unesco-chair-in-global-understanding-for-sustainability/.

Jones, R., T. Jones, and C. Ingram. 2022. Re-Turning Inwards or Opening to the World? Land Use Transitions on Australia's Western Coast. *Geographical Review*, 112(3), 422–443. doi:10.1080/00167428.2020.1856626.

Jones, R., and J. Selwood. 2013. The Politics of Sustainability and Heritage in Two Western Australian Coastal Shack Communities. In *Globalization and New Challenges of Agricultural and Rural Systems: Proceedings of the 21st Colloquium of the CSRS of the IGU*, edited by D.-C. Kim, A. M. Firmino, and Y. Ichikawa, 89–100. Japan: Nagoya University Press.

Liverman, D. M. 2018. Geographic Perspectives on Development Goals: Constructive Engagements and Critical Perspectives on the MDGs and the SDGs. *Dialogues in Human Geography* 8 (2):168–185. doi:10.1177/2043820618780787.

Loffler, R., and E. Steinicke. 2006. Counterurbanization and Its Socioeconomic Effects in High Mountain Areas of the Sierra Nevada. *Mountain Research and Development* 26(1):64–71. doi:10.1659/0276-4741(2006)026[0064:CAISEI]2.0.CO;2.

Long, H., Y. Li, Y. Liu, and X. Zhang. 2013. Population and Settlement Change in China's Countryside: Causes and Consequences. In *The Sustainability of Rural Systems: Global and Local Challenges and Opportunities*, edited by M. Cawley, A. M. D. S. M. Bicalho, and L. Laurens, 123–133. Galway, Ireland: CSRS of the IGU and the Whitaker Institute, National University of Ireland Galway.

Lukić, A., P. R. Kaufmann, L. Valožić, I. Zupanc, M. Cvitanović, D. Pejnović, and I. Žilić. 2022. Croatian Rural Futures in 2030: Four Alternative Scenarios for Postsocialist Countryside in the Newest E.U. Member State. *Geographical Review*, 112(3), 332–352. doi:10.1080/00167428.2020.1871298.

Maher, G. 2013. Attitudes to Brazilian Migrants in Rural Ireland in Conditions of Economic Growth and Decline. In *The Sustainability of Rural Systems: Global and Local Challenges and Opportunities*, edited by M. Cawley, A. M. D. S. M. Bicalho, and L. Laurens, 161–172. Galway, Ireland: CSRS of the IGU and the Whitaker Institute, National University of Ireland Galway.

Moseley, W. G. 2018. Geography and Engagement with U.N. Development Goals: Rethinking Development or Perpetuating the Status Quo? *Dialogues in Human Geography* 8(2):201–205. doi:10.1177/2043820618780791.

Myadar, O., and S. Jackson. 2019. Contradictions of Populism and Resource Extraction: Examining the Intersection of Resource Nationalism and Accumulation by Dispossession in Mongolia. *Annals of the American Association of Geographers* 109(2):361–370. doi:10.1080/24694452.2018.1500233.

Paül, V., and J.-M. Trillo-Santamaría. 2022. The Emerging Mountain Imaginary of the Galician Highlands: A New National Landscape in an Era of Globalization? *Geographical Review*, 112(3), 466–492. doi:10.1080/00167428.2021.1897812.

Robinson, J. 2004. Squaring the Circle? Some Thoughts on the Idea of Sustainable Development. *Ecological Economics* 48(4):369–384. doi:10.1016/j.ecolecon.2003.10.017.

Schmitz, S., and L. Pepe. 2022. Does Local Rural Heritage Still Matter in a Global Urban World? *Geographical Review*, 112(3), 353–370. doi:10.1080/00167428.2021.1890996.

Stedman, R. C. 2006. Understanding Place Attachment among Second Home Owners. *American Behavioral Scientist* 50(2):187–205. doi:10.1177/0002764206290633.

United Nations. 2015. Transforming Our World: The 2030 Agenda for Sustainable Development. A/RES/70/1. https://sustainabledevelopment.un.org/post2015/transformingourworld.

WCED [World Commission on Environment and Development]. 1987. *Our Common Future*. Oxford, UK: Oxford University Press.

Weichelt, R. D., and E. Zeitler. 2022. Second Home Property Ownership and Public-School Funding in Wisconsin's Northwoods. *Geographical Review*, 112(3), 396–421. doi:10.1080/00167428.2020.1855583.

Woods, M. 2005. *Rural Geography*. London, UK: Sage Publications, Ltd.

———. 2011. *Rural*. Abingdon, UK: Routledge.

Zanolin, G., and V. Paül. 2022. Exploring the Sustainability of Wilderness Narratives in Europe. Reflections from Val Grande National Park (Italy). *Geographical Review*, 112(3), 444–465. doi:10.1080/00167428.2020.1869905.

CROATIAN RURAL FUTURES IN 2030: FOUR ALTERNATIVE SCENARIOS FOR POSTSOCIALIST COUNTRYSIDE IN THE NEWEST E.U. MEMBER STATE

ALEKSANDAR LUKIĆ, PETRA RADELJAK KAUFMANN, LUKA VALOŽIĆ, IVAN ZUPANC, MARIN CVITANOVIĆ, DANE PEJNOVIĆ and IVAN ŽILIĆ

ABSTRACT. The academic picture of a globalized European countryside, and particularly of rural areas in postsocialist, new member states of the European Union, is one of huge and increasing complexity, diversity, and uncertainties about the future. The aim of this research is to construct alternative scenarios for rural Croatia in 2030, acknowledging its postsocialist transition as an important framework. Future development scenarios were constructed by integrating quantitative and qualitative approaches. The main methods used were: factor and cluster analysis; Monte Carlo simulation; and Delphi method, involving 37 rural experts in two rounds of written questionnaires. Four scenarios were developed: Rural Renaissance, Shift, Road to Nowhere, and Growth without Development. These scenarios provide a set of well-documented and reasonable assumptions to aid in thinking about possible future paths for the Croatian countryside, while at the same time allowing for the discussion of rural development paradigms.

The overwhelming academic picture of the European countryside has been, for some time now, one of great and increasing complexity, diversity, and heterogeneity (Halfacree 2006; Cloke 2006; OECD 2006; Rienks 2008; Copus et al. 2011; Woods et al. 2015). In general, the contributing factors are well-researched and related to the impacts of interconnectedness, the importance of linkages and flows, and the changing relational aspects of rural and urban (Copus et al. 2011; Woods et al. 2015; Dax and Copus 2016). As global linkages and connections are eased by technological changes and influenced through political frameworks of deregulation, local urban-rural interactions are taken over by a web of interactions on multiple spatial levels (local, regional, national, European, and global) (Copus et al. 2011), increasing rural heterogeneity.

Furthermore, stronger integration of rural areas into wider spatial systems has simultaneously brought a higher level of dependence on numerous external factors—economic, political, social, cultural, and environmental—to rural Europe, raising uncertainties for its future development. For example, exposure to external threats from changing terms of trade has proven to be a very important issue in rural Europe (Rizov 2006), as has climate change (see the EDORA Final Report 2011). Academic response to the increased level of uncertainties has raised interest in applying the scenario method in investigating possible ways forward for diversified European rural areas (for example, ATEAM; SCENAR2020 and SCENAR2030; EURURALIS; or ET2050).

New member states (NMS) from Central, Eastern, and Southeast Europe.[1] have, along with globalization and/or as a part of it, undergone major transformations in the last 30 years: political and economic transition and accession to the E.U. These processes have brought additional levels of complexity to discussing possible pathways into the future, especially for the countryside. The NMS have been more or less present in a series of E.U.-wide research projects concerning modeling and scenario planning in rural areas. However, unlike E.U.-15 countries, for which numerous national scenarios, including rural areas, have been produced, scenarios are rare for the NMS (see Bański 2017). This paper contributes to filling recognized gaps by presenting the results of four alternative scenarios for rural Croatia in 2030, in the context of rural development, transition, and accession to the E.U.

Theoretical Background

Changing Paradigms of Rural Development

Our hypothesis is that the alternative future prospects of rural areas are related to differentiated outcomes of the interactions between the unique setup of natural and social capital (endogenous resources) available in the region on one hand, and the exogenous conditions, actors, and processes on the other. This follows from the change of modernization's paradigm to a new concept of rural development (Ray 1999; Van der Ploeg et al. 2000; Van der Ploeg and Marsden 2008; Woods 2011). Instead of the linear development trajectory of modernization theory, which mainly focuses on progress in agriculture, industrialization, and urbanization, the paradigm of new rural development is characterized by a web of nonlinear multitude or, as Jan Douwe Van der Ploeg et al. (2000) argued in their seminal paper at the turn of the twenty-first century, by its "multi-level, multi-actor and multi-facetted nature." In their later work, Van der Ploeg and Marsden (2008) recognized six components of the (new) web of rural development: endogeneity, novelty, market governance, new institutional frameworks, sustainability, and social capital.

Woods (2011) further summarizes the differences between the new rural development paradigm and the modernization paradigm by acknowledging its distinguishing features: endogenous development, bottom-up innovation, territorially based integrated development, social capital, sustainable development, information infrastructure, consumption, small-scale niche industries, valorization of tradition, and local embeddedness. Ray (2006) in particular has further explored the idea of the multiactor character of new rural development.

He argues that besides local actors who mobilize cultural capital—that is, "territorial intellectual property or place-specific factors of production" (Ray 2006, 283)—through bottom-up development, various actors from the "extralocal environment" are potential supporters of the local population in their development strategies. He expressed these "various manifestations of the extralocal" by adding the prefix "neo" to endogenous development, thus coining the term "neoendogenous rural development." In the E.U. context, the most visible pragmatic expression of the new paradigm of rural development is usually attached to LEADER Programme. Although its advantages for rural areas have been recognized, there has also been criticism, mostly regarding the lack of real participation, inclusiveness, innovation, and capacity to address problems of structural disadvantages, which are also related to the limited and insufficient financial resources of the CAP budget (Marquardt et al. 2010; Dax et al. 2016; Dax and Copus 2016; Lukić and Obad 2016).

SCENARIO METHOD AND RURAL AREAS

A further aspect of (new) rural development theory is related to the discussion about the future of rural areas. Van der Ploeg et al. (2000) argued that the concept of rural development was primarily "a heuristic instrument ... (that) ... represents a search for new futures ... " The scenario as "[a] description of a future situation and the course of events that allows us to move from the original situation to the future situation" (Godet and Roubelat 1996, 166) was chosen to reflect this aspect of rural development in our research.

The origins of the scenario method can be traced to the period during and after the Second World War, when it was used for military purposes (Schoemaker 1993). The scenario method further developed through its use in public policy and corporate/business planning, as well as in the scope of urban and regional planning, primarily in the United States and France (Godet and Roubelat 1996; Bradfield et al. 2005). Although the interest in scenarios has fluctuated somewhat over the years, it increased in the 1990s and especially 2000s (Ringland 2006; Varum and Melo 2010). Scenario construction leads to scenario analysis and scenario planning where, in essence, by examining several alternatives of how the future might unfold and comparing the potential consequences of different future contexts, one "can make decisions which are more resilient to the throes of tomorrow" (Shearer 2005, 68).

Since the 1990s, there have been a number of important research projects/programs exploring European rural futures, such as Ground for Choices, ATEAM, ACCELERATES, SCENAR2020 and SCENAR2030, EURURALIS, ESPON, FARO-EU, EDORA, TIPTAP, ET2050, and VOLANTE. For example, within the European Development Opportunities for Rural Areas (EDORA) Project, opportunities and constraints for different kinds of rural areas were considered over the course of two decades. Narrative scenarios were formed based on the relationship between two "external" variables, namely climate change (and responses) and economic governance (EDORA Final Report 2011).

Some of the recent topics in rural research explored via the scenario method include studies of complex rural development changes (Bański 2017), land use and land-cover change (Holman et al. 2017; Kindu et al. 2018), and the future of farming areas (Zagaria et al. 2017). The scenario method is regarded as a participatory and strategic planning tool that provides a sound basis for policy and decision making, and facilitates communication among stakeholders (Soliva et al. 2008; Gullino et al. 2018). Different scenario typologies classify scenarios in relation to two basic questions to which scenarios respond: regarding the development that could happen and the development that should happen—that is, scenarios that are explorative or normative in nature (Radeljak Kaufmann 2016). A plethora of methods and techniques used in developing scenarios encompass modeling approaches (Holman et al. 2017; Zagaria et al. 2017; Kindu et al. 2018), Delphi-based surveys (Trammell et al. 2018), participatory workshops or stakeholder meetings (Soliva et al. 2008; Holman et al. 2017), focus groups (Gullino et al. 2018), and rural stakeholder/expert interviews (Soliva et al. 2008; Rawluk and Curtis 2017; Gullino et al. 2018). Various methods and sources are often combined (Priess et al. 2018). Interactive scenarios result from creative processes and interactions among people (experts or users) and endogenous or exogenous models, either during the formation of scenarios or during their use (Gordon and Glenn 2018).

Setting the Scene: Challenges for the Rural Areas in the New Member States and Croatia

A quarter of a century ago, the future looked bright and promising for countries in Central, Eastern, and Southeast Europe. Dreams of unifying Europe after the Cold War spurred optimism. Liberal democracy, freedom, civic society, and a market economy were mostly warmly welcomed (Sokol 2001). Although accession to the E.U. was still far away, the importance of the E.U.'s Common Agricultural Policy (CAP) was also raising hopes that rural areas, after the turmoil of collectivization, negligence, and ideological subordination in industrially oriented state-socialist societies, would be given a new push. Fast forward to 2020 and the situation in most of the rural regions in the NMS is certainly much more complex and worrisome than expected. The political and economic transition and the effects of E.U. accession have had both positive and negative effects (Spoor et al. 2014; Swain 2016; Cianetti et al. 2018).

On the positive side, the former significant differences between the NMS and the EU-15 in GDP per capita, economic activity, or long-term unemployment seem to have decreased (Sokol 2001; Ezcurra et al. 2007; Gorton et al. 2009). Furthermore, in the majority of the NMS the financial support to agriculture and rural areas has increased and accession to the E.U. is generally perceived as having a positive impact (Rizov 2006; Swain 2016).

On the negative side, some serious challenges for the future development of rural areas in the NMS have been noted: persisting and/or deepening regional disparities, polarization, and peripherization of rural areas (Sokol 2001; Ezcurra et al. 2007; Macours and Swinnen 2008; Kovács 2010; Cosier et al. 2014; Spoor et al. 2014; Swain 2016; Páthy 2017; Loewen and Schulz 2019); inadequacy of agricultural transformations related to the E.U.'s Common Agricultural Policy (CAP) (Rizov 2006; Macours and Swinnen 2008; Kovács 2010; Swain 2016); and rural shrinkage (long-term depopulation and youth out-migration) (ESPON ESCAPE 2019).

National specificities notwithstanding, especially direct and indirect consequences of the Croatian War of Independence (1991–1996), Croatia shared many of the analyzed features of postsocialist NMS. Between 2001 and 2017 Croatia lost (officially) 331,967 inhabitants (7.5 percent). In that period, the demography of Croatia was characterized by a decline in total population, continuous natural depopulation, increased aging of the population, imbalanced age structure of the population, and a positive net foreign-migration balance (though this has been negative since 2009). At the moment, the Croatian population ranks among the top 15 oldest world populations, and the share of older people in the total population is constantly increasing (Čipin et al. 2014).

Rural areas in Croatia, which encompass around 90 percent of the total land area and around 49 percent of the total population, bear many of the aforementioned national negative trends, but with even more undesirable tendencies. The typology of rural areas developed in this scenario study revealed that almost one-fifth of the rural territory belongs to areas with critical demographic characteristics and economic limitations in development, while an additional 18.6 percent of territory is considered to be predominantly agricultural areas, with significant unemployment levels (Table 1).

Thus, almost 40 percent of the rural territory is burdened by socioeconomic decline and/or rising unemployment rates. Although rural depopulation and marginalization is a legacy process in Croatia, the Croatian War of Independence and numerous economic difficulties in the transition period in the 1990s intensified external migration, deepened polarization between the wider Zagreb region and certain touristed coastal areas, and most of the Croatian countryside. Rural youth out-migration was increased after Croatia's accession to the E.U. Regional inequalities and the urban/rural divide have also increased. Prior to E.U. accession, agriculture in Croatia was considered weak and unable to compete in the E.U. (Franić et al. 2014), while the consequences of the CAP have yet to be seen.

Table 1—Share of Territory (T) and Population (P) for Clusters/types of Rural Areas of Croatia in 2017 and the Four Scenarios

Type of rural area	Typology in 2017		Rural renaissance		Shift		Road to nowhere		Growth without development	
	T (%)	P (%)	T (%)	P (%)	T (%)	P (%)	T (%)	P (%)	T (%)	P (%)
C1—Predominantly agricultural areas, with significant unemployment levels	18.6	18.3	19.8	19.5	19.4	19.0	19.2	19.3	24.7	23.6
C2—Heavily agricultural areas	20.0	16.4	18.2	15.3	19.2	15.4	5.1	4.6	11.4	10.4
C3—Strongly urbanized and demographically more dynamic areas, with lower significance of agriculture	6.2	19.2	4.0	9.2	4.5	15.0	5.4	18.7	2.8	10.7
C4—Touristic, demographically older areas	10.6	7.1	13.6	11.3	14.7	13.9	9.3	7.2	11.9	9.0
C5—Economically diversified areas, demographically and economically less threatened	24.7	30.6	23.2	36.0	21.2	28.1	33.4	33.2	26.3	37.1
C6—Areas with critical demographic characteristics and economic limitations in development	19.8	7.1	21.1	7.4	20.9	7.4	27.5	15.7	22.8	7.8

Methods in Explorative Scenario Development for Rural Croatia until 2030

The process of constructing future development scenarios for rural Croatia until 2030 included multiple steps. The selection of quantitative steps in scenario development was partially based on a study of rural England, where the results of factor and cluster analyses served to create a model to simulate rural dynamics and, via a Monte Carlo process, to develop scenarios (Foa and Howard 2006; Lowe and Ward 2009).

In our study, rural Croatia was defined as incorporating rural and mixed rural/urban areas, whereby the spatial units in the analysis were local government units (LAU2), excluding urban settlements with a population over 5,000 in 2011. In the first step, a total of 43 variables (derived from the official statistical sources and project analyses) as indicators of development trends in rural Croatia were used in factor analysis. The initial factor loadings were calculated using the principal component method, while the rotation method used was Oblimin with Kaiser Normalization. Based on the proportion of variance explained by each of the seven resulting factors, a total of 15 leading variables were identified as key for the development of rural areas:

1. Proportion of persons employed in accommodation and food-service activities in the settlement where they work, 2011 (empl_accomm)
2. Proportion of the population with property income in the total population, 2011 (pop_prop)
3. Coefficient of tourist functionality (number of tourist beds per inhabitant), 2011 (coeff_tour)
4. Proportion of persons employed in the tertiary sector of the economy in the settlement where they work (excluding real estate, information, and communication), 2011 (empl_serv)
5. Proportion of the population aged 15 and over with a university degree, 2011 (pop_uni)
6. Proportion of the population aged 0 to 19, 2011 (age0_19)
7. Proportion of the population aged 60 and over, 2011 (age_60ov)
8. Proportion of nonfamily households (singles) to the total number of private households, 2011 (singl_priv)
9. Natural population growth rate, 2001–2011 (nat_growth)
10. Proportion of persons employed in the primary sector of the economy in the settlement where they work, 2011 (empl_prim)
11. Proportion of agricultural holdings to total number of households, 2011 (agric_hold)
12. Proportion of unemployed persons to economically active, 2011 (unemp)
13. The average size of the ARKOD arable land unit (ha), 2015 (size_unit)
14. Proportion of built-up area, 2012 (built_up)
15. Proportion of in-migrants in a settlement from another settlement of the same local government units to total population, 2011 (inmigr_lgu)

Table 2—Mean Values of Variables in Six Clusters (Types of Rural Areas) in Typology in 2017, Rural Renaissance and Road to Nowhere Scenarios

TYPE	EMPL_ACCOMM	POP_PROP	COEFF_TOUR	EMPL_SERV	POP_UNI	AGE0_19	AGE_60OV	SINGL_PRIV	NAT_GROWTH	EMPL_PRIM	AGRIC_HOLD	UNEMP	SIZE_UNIT	BUILT_UP	INMIGR_LGU
Typology in 2017															
C1	4.7	0.2	3.7	21.3	5.5	23.0	23.4	23.3	−36.7	24.6	17.3	25.4	2.4	2.5	6.8
C2	2.9	0.3	1.5	15.3	4.9	22.1	25.5	24.1	−62.6	49.4	38.6	18.3	1.1	2.3	8.0
C3	12.2	0.9	83.7	38.7	14.0	20.8	23.2	20.9	0.7	3.5	7.0	14.8	0.5	8.6	12.1
C4	24.6	3.7	249.3	43.7	13.1	17.5	31.5	29.2	−55.3	13.3	17.1	14.0	0.3	2.5	5.6
C5	6.5	0.2	15.3	29.0	7.9	22.0	23.7	21.4	−27.9	8.1	18.1	15.9	0.4	3.4	8.2
C6	5.2	0.1	4.8	24.0	6.3	16.5	35.3	33.1	−139.5	14.1	14.6	25.5	0.5	1.0	9.4
Rural Renaisannce Scenario															
C1	8.8	0.3	5.6	23.9	8.6	21.4	22.3	21.9	−38.0	23.6	19.5	20.5	3.1	3.1	4.8
C2	4.4	0.4	3.2	18.5	8.4	22.5	23.0	22.3	−69.0	47.7	40.1	14.2	1.9	3.5	5.7
C3	19.9	0.8	70.6	51.4	16.8	24.4	21.6	16.9	41.8	2.8	4.8	9.8	0.5	8.1	26.8
C4	28.3	3.4	223.1	54.2	18.3	18.7	30.5	26.4	−66.7	11.9	13.2	8.7	0.3	4.7	4.4
C5	11.8	0.7	32.3	40.9	12.7	23.7	23.1	18.5	−21.9	5.8	16.6	10.2	0.6	7.5	6.8
C6	8.7	0.3	6.6	29.8	8.7	16.2	33.3	32.4	−142.9	10.9	18.1	19.3	0.7	1.3	5.4
Road to Nowhere scenario															
C1	3.9	0.8	3.3	22.2	5.0	15.1	28.7	27.6	−32.6	17.9	12.6	28.4	2.8	2.7	7.3
C2	2.7	0.5	1.4	13.2	3.7	14.9	33.9	27.3	−74.3	38.7	37.3	16.6	1.1	2.3	12.4
C3	12.2	1.8	80.8	42.6	13.7	16.2	27.1	25.4	−7.1	5.0	8.3	16.2	0.6	9.8	17.5
C4	25.8	8.8	358.8	51.9	13.7	12.4	37.1	36.3	−44.0	12.7	15.7	17.1	0.3	2.8	10.5
C5	7.8	0.3	28.4	31.5	7.5	16.0	28.7	26.7	−35.2	8.2	15.2	18.3	0.5	3.2	13.0
C6	2.8	0.4	2.0	20.2	4.7	14.0	36.7	33.9	−82.3	27.8	22.3	27.7	1.5	1.9	10.2

Second, a cluster analysis was conducted where six different types of rural areas in Croatia were recognized (Tables 1 and 2). Ward's clustering method was used (squared Euclidean distance, values standardized by variable between −1 and 1).

The following step in our scenario construction included a Delphi study, consisting of two rounds of written questionnaires. The Delphi panel encompassed academics with rural expertise. The choice of panelists was based on papers published by potential members, their scientific discipline, and regional distribution. Out of 57 researchers who were initially contacted via e-mail, 37 participated in the first Delphi round, while 13 participated in the second round. The following disciplines were represented: agronomy (3), agricultural economics (6), anthropology (2), architecture and urbanism (1), demography (3), economics (3), ethnology (1), forestry and environmental protection (1), geography (9), spatial economics (2), social work (1), and sociology (5). The time frame was June/July 2017 for the first Delphi round, and August/September 2017 for the second round. The first questionnaire was disseminated via an online service and the second via e-mail. In the first round the panel considered possible future development of each of the 15 key variables (through numerical estimates and additional explanations). They also had an option to describe other factors deemed important for the future development of Croatian rural areas, either in a national or international context. For the sake of clarity, the questionnaire was prechecked by two experts who later participated in the Delphi panel; their comments helped improve the final questionnaire structure.

The analysis of the results was both quantitative and qualitative. Numerical estimates were used as input data for Monte Carlo simulations, resulting in the most likely new types of rural areas (clusters) in the final year. The qualitative analysis of responses to open-ended questions was conducted with regard to the possible future developments of listed variables, including the specific characteristics of rural area types, and other factors, both endogenous and exogenous. These results served as input in structuring the scenarios.

Scenarios were developed using the scenario axes technique/2 x 2 matrix, which Bishop et al. (2007) refer to as the Royal Dutch Shell/Global Business Network matrix approach. The approach includes identifying the two most important and most uncertain factors for future development. The aforementioned factors are represented by two axes, whereby opposite axes sections stand for extreme values of each factor. The resulting four quadrants set the frame for possible future developments within four different scenarios.

In our study, two axes of key uncertainties were recognized based on the qualitative analysis of the first Delphi round. We also used the results of a questionnaire survey among 59 rural stakeholders who were participants at the First Croatian Rural Parliament. The selected key uncertainties were demographic state and processes, and a combination of innovation—such as problem solving, economic, and technological development—and actor networking (Figure 1).

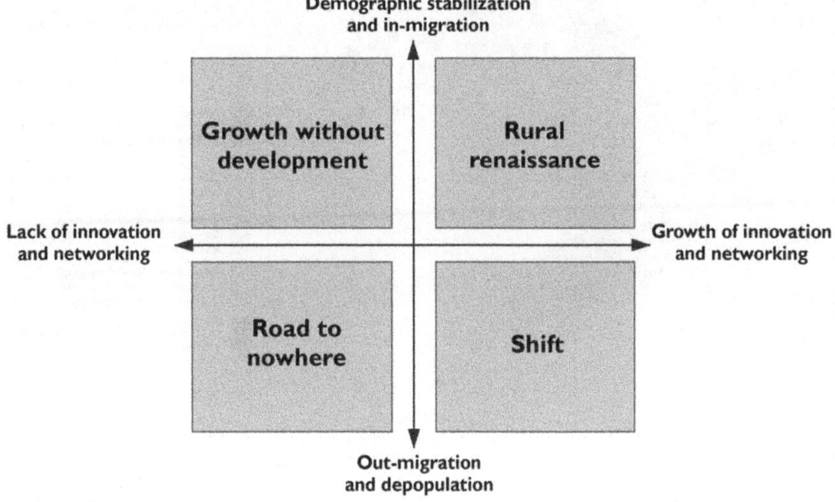

FIG. 1—Scenarios on Croatian rural futures until 2030—scenario axes.

Numerical estimates for 15 key variables and different types of rural areas were organized along the two axes by the expert team, which served as the backbone for scenarios and an input into Monte Carlo simulations, and ultimately led to four new rural typologies (for example, Figures 2 and 3). Scenario narratives were based primarily on the qualitative analysis of the Delphi survey responses. In the second

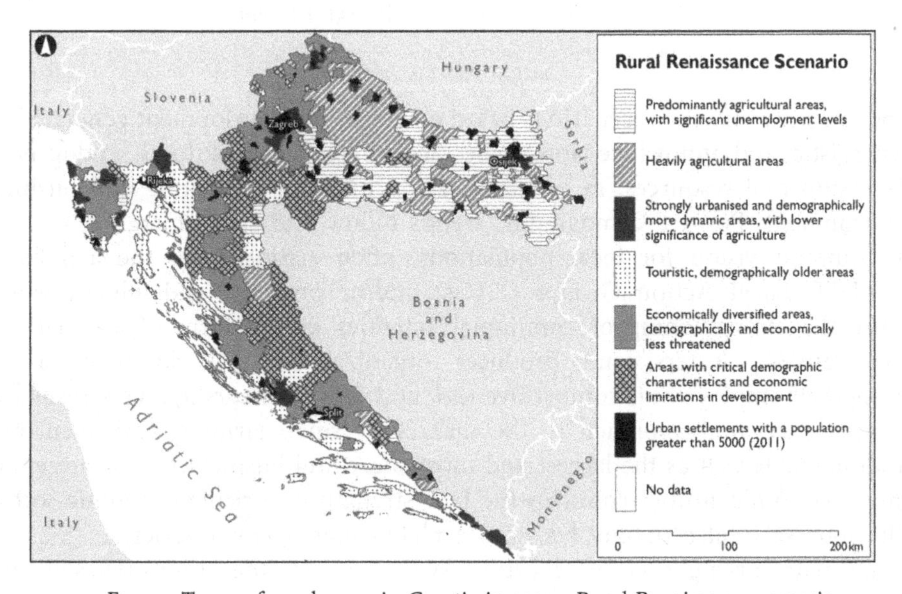

FIG. 2—Types of rural areas in Croatia in 2030—Rural Renaissance scenario.

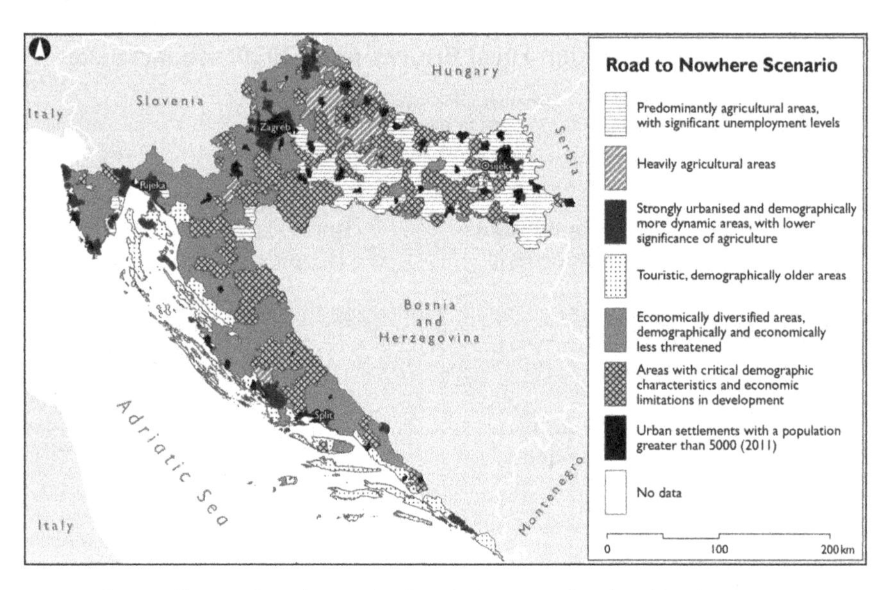

FIG. 3—Types of rural areas in Croatia in 2030—Road to Nowhere scenario.

Delphi questionnaire, the participants gave their comments on scenario narratives, including the consistency-check and the results of numerical simulations. The constructed scenarios are explorative scenarios, based on the analysis of current trends and exploration of possible future developments.

SCENARIOS ON CROATIAN RURAL FUTURES

RURAL RENAISSANCE

The Rural Renaissance scenario is based on economic development generated by synergistic and innovative approaches to the best use of local, endogenous developmental resources, in an encouraging institutional setting and business environment. Positive demographic trends follow and start emerging by 2030. A common vision for local populations, often arising from the actions of LEADER Local Action Groups (LAGs) and/or proactive individuals, would cause the strengthening of communal activities in the form of associations, cooperatives, clusters, and producer organizations that, in turn, would strengthen employment, competitiveness, and entrepreneurship. This results in a more successful approach to the market by small farmers, craftsmen, and tradesmen, as well as the largest and most successful businesses. The success of small economic entities confirms the true strength of a rising economic sector. The multisectoral economy becomes a reality that opens a variety of jobs for people with varying levels of education. With respect to the various types of rural areas, multisectoral economy encompasses market-oriented agricultural production of traditional crops, new high-income crops and organic agriculture, the combination of agriculture and tourism, and the development of many other selective forms of tourism based on natural and cultural heritage. In the vicinity of urban centers, there is often development of innovative and technologically advanced industries. Decisions regarding the use of the most important local development resources of rural communities are made with the consent of local, regional, and national governance levels. This represents the first step toward the successful decentralization of the institutional framework, which is experiencing positive change under the influence of an increasing number of different actors on one hand, and the demands that come with obtaining money from E.U. funds on the other. Administrative obstacles regarding entrepreneurship, as well as the uncertainty of the legal framework have been alleviated. More significant influence of funding from Common Agricultural Policy (CAP) and other relevant E. U. funds is expected in the new budget cycle after 2020, with the strengthening of absorption power on the local level. As a result of the aforementioned, rural development stakeholders would not allow a return to "business as usual," that is, passive observation of the progression of negative processes. In such an atmosphere, the inefficient and uncoordinated state and/or arbitrariness of local leaders is gradually replaced by positive engagement of local stakeholders and various external factors in a relationship that transitions from paternalistic to partner relations. This enables better quality of planning and the development of communal and social infrastructure that is essential to the quality of life. Demographic measures are designed on the basis of immigration and redistributive demographic policies, with a focus on young families; they are more than just maternity benefits and housing support funds, and are primarily oriented toward supporting the working and social environments. A growing number of young urban residents who have just finished their education have decide to move to rural areas for new opportunities for employment, to start their own businesses, for the rural lifestyle, and/or due to insecurity of employment in urban centers. As a result of the lengthening of the tourism season, seasonal workers also come to work in tourist destinations and reside in nearby rural areas, alongside immigrants from war-afflicted areas. The influence of potential E.U. enlargement to Croatia's neighboring countries and general geopolitical stability could contribute to the strengthening of rural development in Croatia, both thanks to increased trade and the stabilization of cross-border relations.

SHIFT

The Shift scenario portrays the future of a rural Croatia in which a new approach to solving key rural problems is applied, but with a continued negative demographic situation, emigration, and depopulation as dominant trends. The described positive changes in the Rural Renaissance scenario also take place in this scenario, but they are slower and less intense, which does not result in the cessation of negative demographic trends by the end of the investigated period. An important innovative turnabout regarding the "activation" of the older population, however, does takes place, which is a change from this group's neglected role in current development processes. They are recognized as an important factor of economic development due to their skills and traditional knowledge, as well as their ownership of land and property. In this scenario, in contrast to multisectoral economic development in the Rural Renaissance scenario, the advancement of agriculture is the key to development. This is the consequence of innovative approaches to solving some key contemporary problems, such as land ownership relations and agricultural land management. When attracting funds, the emphasis has primarily been on the Rural Development Programme within CAP and strengthening of agricultural competition. Positive effects in agriculture are evident: holdings grow in size, the availability of various forms of professional education grows, the position of small agricultural producers improves due to innovative links to urban markets, and agriculture becomes increasingly multifunctional. The other strong pillar of economic development is tourism. Owing to innovative approaches in the development of selective forms of tourism, areas near coastal tourism hubs enjoy strong tourism development. There is a growing need for workers in both economic branches, and demographic policy measures, despite being planned and carried out, have not yet given results. The possibility of revitalization is found in the incentive for redistribution of population from other parts of Croatia; however, the "import" of foreign workers from neighboring countries is used as the main compensatory economic measure. This then creates a permanent state of uncertainty in regard to the unstable geopolitical situation in Southeast Europe and the uncertain destiny of further E.U. enlargement.

ROAD TO NOWHERE

The Road to Nowhere scenario is determined by emigration and depopulation, the lack of innovation, and discordant approaches on the part of a growing number of actors. There is a lack of development incentive on the local level, primarily due to further weakening of human and social capital, and dependence of the economic sector on large entities, primarily in agriculture and tourism. The strong outflow of the young, educated population continues, which brings about an irreversible loss of new knowledge and new technological approaches in agriculture and the economy. The lack of human resources, especially youth, becomes the main factor of lagging behind in development.

Under such conditions, the institutional framework and top-down actors continue with paternalistic relations that result in the continuation of policies of political and fiscal centralization, with emphasis on strong local, primarily political, actors. The expected positive influences of the Rural Development

Programme and E.U. funds are lacking, with a few exceptions, due to weak usage, caused by disharmony on all levels of administration and a lack of experts. Furthermore, the funding that is absorbed does not have a wider territorial effect, because it is hindered by numerous bureaucratic obstacles (frequent changes to the legal framework, stifling tax structure, and trade difficulties). Demographic policy remains on the level of maternity benefits and some local measures, without a systematic approach.

Increasing unemployment is a key factor in further depopulation. Under conditions of a shrinking local job market and vanishing social amenities, it is nearly impossible to find seasonal workers for agriculture or tourism. This encourages unplanned and unselective import of predominantly unqualified workers. Along with the aforementioned processes, local development resources, primarily natural—freshwater, forests, biomass, stone, soil—become fundamental, but nameless, sources of income, mostly in the hands of foreign investors. Local identity—natural and cultural heritage—is degraded due to an exclusive market orientation. The aspiration level of the remaining youth falls. Systematic, balanced, and overall development becomes impossible under conditions of increasing peripheralization of rural Croatia. Border areas are especially threatened due to unsolved border disputes between Croatia and neighboring countries. In this sense, geopolitical issues and the consequences of the migrant crisis, if it continues, can have a strong negative influence on rural Croatia.

GROWTH WITHOUT DEVELOPMENT

The Growth without Development scenario describes a future of rural Croatia in which stagnation or population growth is achieved by 2030, depending on the type of rural area. However, this happens in an environment that lacks innovative approaches to solving key development problems and does not advance the level of coordination and networking of key actors and stakeholders. Consequentially, in contrast to the similarly demographically positive Rural Renaissance Scenario, this sort of quantitative trend is not reflected on the qualitative level, as numeric simulation shows that the share of youth will continue to reduce along with a continual increase in the share of aged population. This situation reveals that in-migration has primarily been a consequence of external trends, of which the most important are further tourism and an increase in the number of second homes in touristic areas of Croatia that become seasonal/permanent places of residence for an increasing number of retired Europeans. For retirees, Croatia is an increasingly attractive, secure place with a favorable climate. Furthermore, after the first strong wave of emigration from eastern Croatia, the demographic situation there stabilizes, given that the younger and the more ambitious have left, and the remaining population is more passive, and also because owning agricultural land offers at least some kind of revenue.

The development of mass tourism at the seaside is the dominant trend in this scenario, which demands a large, less educated workforce, so immigration is on the rise but is very unselective and is the result of business interests of investors and owners of tourist facilities. In the European context, this scenario partially hints at the idea of Croatia as a "European playground," highlighting the country's weak importance in terms of the production sector, and its significance in terms of leisure and recreation—especially due to the growth in the number of affluent retirees. Geopolitical processes in the (wider) neighborhood are of key importance, because any sort of instability repels the main source of income and endangers economic development.

SPATIAL IMPACTS

Spatial impacts of explorative scenarios were visualized through four scenario-specific typologies. Here we explore the examples of two scenarios: Rural Renaissance (Figure 2) and Road to Nowhere (Figure 3). The mean values for each of the 15 variables in different types of rural areas (Table 2) indicate the difference in trends in the Rural Renaissance and Road to Nowhere scenarios with regard to the original typology.

The Rural Renaissance scenario resulted in a significant increase in the number, area, and population in the cluster of touristic, demographically older areas (10.6 percent to 13.5 percent of territory, 7.1 percent to 11.3 percent of population). Owing to the development of selective forms of tourism in the hinterland, parts of mountainous Croatia near Kvarner, the island Krk, Istria, and Šibenik Zagora became a part of this type. Furthermore, parts of Slavonia, in the wider Osijek region, have undergone positive changes in terms of development of agriculture and general employment level, which can be observed in their transition from predominantly agricultural areas, with significant unemployment levels, into heavily agricultural areas.

The Road to Nowhere scenario brings the most visible spatial changes in the distribution of types of rural areas. Areas with critical demographic characteristics and economic limitations in development had especially large growth in the number of spatial units and, consequentially, in the shares of surface area and population (increase were experienced from 19.8 percent to 27.5 percent of territory, and from 7.1 percent to 15.7 of population.) The second-expressed spatial change is the strong reduction of heavily agricultural areas (decrease from 20.0 percent to 5.1 percent of territory, and from 16.4 percent to 4.6 percent of population), which is a direct consequence of the aforementioned spread of areas with critical demographic characteristics and economic limitations in development.

Discussion and Conclusion

Our study was guided by the idea that "the theory of rural development does not deal with the world as it is—it is about the ways in which agriculture and landscape could be redesigned" (Van der Ploeg et al. 2000, 396). By developing four alternative scenarios with different demographic, economic, social, and spatial outcomes, we have offered a set of well-documented and reasonable assumptions that should help us to reflect on possible future paths for Croatian rural areas.

Numerical simulations have shown that alternative scenarios imply a change in the spatial distribution and size of the territory, as well as in the number of inhabitants in a given rural area type compared to the current typology (Figs. 2 and 3). Although some previous research has also warned that future studies need more integrative and robust methods to use the concept of a heterogeneous countryside as a starting point (Foa and Howard 2006; Lowe and Ward 2009), scenario exercises usually consider ex-post spatial impacts of developed scenarios in different types of rural areas. In our study, the diversity of rural areas was the starting point and was considered simultaneously with the development of scenario narratives in an attempt to address identified challenges.

Furthermore, with regard to rural development theories, we would argue that the alternative scenarios developed for rural Croatia in 2030 consist of components of both the modernization paradigm and the new rural development paradigm. This particularly concerns recognized factors that could influence future rural development in Croatia: endogenous development vs. inward investments, multilevel governance vs. paternalistic, centralized state, sustainable development vs. exploitation and control of nature, consumption vs. production, multisectoral economy vs. sectoral dependency (tourism), and peripheralization vs. balanced regional development. The Rural Renaissance and Road to Nowhere scenarios seem to correspond to the tension between the paradigms. Rural Renaissance is the most positive picture of the Croatian countryside in 2030, and it is the multitude of different actors from different sectors of society—institutions, enterprises, citizens, individuals—focusing on territorial resources and working in vertical and horizontal partnerships, that bring progress, reflecting the new paradigm of rural development, especially its neoendogenous aspect. In Road to Nowhere, however, the opposite is true; the lack of social and entrepreneurial energy due to weak human and social capital makes dependence on foreign investment and top-down development inevitable. Instead of sustainable development, which is emblematic of the new paradigm of rural development, Road to Nowhere brings further exploitation of natural resources, sectoral development, and dependence on imported financial capital, features that are related to the instruments associated with the unsuccessful application of the modernization paradigm. On the other hand, the two other scenarios are arguably much more mixed in terms of their components in relation to the rural development paradigms. For example, while the postproductivist, consumption countryside is usually considered as a domain of the new rural development, the Growth without Development scenario depends strongly on tourism and the growing second-home sector in rural Croatia in 2030, while there is no significant effort in the areas of integrated territorial development, networking, multilevel governance, and innovative approaches, important aspects of new approaches to rural development. Many of the Delphi respondents considered the aforementioned paradigmatically mixed two scenarios as more realistic versions of the rural futures. This could be an interesting proposal for future research on the currently dualistic and separate view of old and new theories of rural development.

Moreover, the scenarios emphasized the role of regional development for the future pathways for rural areas. In the Road to Nowhere scenario, the term "peripheralization" was frequently used by experts in Delphi research to describe the process of deepening disparities between the capital city and its urban region on the one hand and lagging rural areas on the other. The experts shared the view that center-periphery relations are becoming one of the main obstacles to a systematic and balanced development of rural Croatia. Similarly, polarization processes between urban/metropolitan areas and rural areas have been identified in many NMS as negative impacts on intraregional cohesion and rural development (Binelli and Loveless 2016; Gorton et al. 2009; Kisiała and Suszyńska 2017; Brambert and Kiniorska 2018; Pociūtė-Sereikienė and Kriaučiūnas 2018). They appear to be of particular importance even in the economically better-off NMS such as Slovenia (Cosier et al. 2014), where studies on rural development also suggest that the strengthening of urban-rural links in terms of improving the employment and residential conditions, infrastructure, and access to services could bring mutual benefits to regional development of the county as a whole (Perpar 2014).

Finally, the question could be raised about the possibility that, due to the diversity of rural areas, there could be a mixture of scenarios in each rural environment, leading to potential conflicts of interest among them (Guštin and Potočnik Slavič 2020). Our modeling of the different scenarios, as illustrated by the changes in the case of Rural Renaissance and Road to Nowhere (Figs. 2 and 3, Table 2), statistically express the most likely outcome of the selected scenario at the spatial level of LAU2 (municipality). However, we do not claim that only one scenario is possible within a particular type of rural area, especially in municipalities or settlements belonging to a particular type. A different approach to the use of typology in the study of rural heterogeneity was taken by Woods et al. (2015), who developed a typology of rural responses to the challenges of globalization. They identified eight different types of regional responses that can occur in different rural settings (for example, Relocalizers, Global Conservators, and Resource-providers). However, in that study the comprehensive view of the territorial typology—inclusion of all territorial units in the spatial framework studied—is not present since the methodology is based on selected case studies. In future research, the combination of these two typologies, typology of rural areas and typology of rural responses, in a newly developed and methodologically robust framework could be a step toward new perspectives for understanding current and future processes in diversified rural areas.

Funding

This work has been supported by the Croatian Science Foundation under the project number 4513 (project title: Croatian Rural Areas: Scenario-based Approach to Discuss Planning and Development). Any opinions, findings, and conclusions or recommendations expressed in this material are those of the author(s) and do not necessarily reflect the views of the Croatian Science Foundation.

Note

[1] In this paper the term "new member states" (NMS) is used to denote ex-socialist countries of Central, Eastern, and Southeast Europe that entered the E.U. in the last three waves of enlargement: in 2004 (the Czech Republic, Estonia, Hungary, Latvia, Lithuania, Poland, Slovakia, and Slovenia), 2007 (Bulgaria and Romania), and 2013 (Croatia). Although Malta and Cyprus joined the E.U. in 2007, they are not considered in this paper.

Orcid

Aleksandar Lukić http://orcid.org/0000-0002-7884-5818
Petra Radeljak Kaufmann http://orcid.org/0000-0003-2738-4035
Luka Valožić http://orcid.org/0000-0001-5696-0367
Marin Cvitanović http://orcid.org/0000-0002-3741-0332

References

Bański, J. 2017. The Future of Rural Poland: The Main Trends and Possible Scenarios. *Eastern European Countryside* 23 (1):71–102. doi:10.1515/eec-2017-0004.

Binelli, C., and M. Loveless. 2016. The Urban–Rural Divide. *Economics of Transition and Institutional Change* 24 (2):211–231. doi:10.1111/ecot.12087.

Bishop, P., A. Hines, and T. Collins. 2007. The Current State of Scenario Aevelopment: An Overview of Techniques. *Foresight* 9 (1):5–25. doi:10.1108/14636680710727516.

Bradfield, R., G. Wright, G. Burt, G. Cairns, and K. van der Heijden. 2005. The Origins and Evolution of Scenario Techniques in Long Range Business Planning. *Futures* 37 (8):795–812. doi:10.1016/j.futures.2005.01.003.

Brambert, P., and I. Kiniorska. 2018. Changes in the Standard of Living in Rural Population of Poland in the Period of the EU Membership. *European Countryside* 10 (2):263–279. doi:10.2478/euco-2018-0016.

Cianetti, L., J. Dawson, and S. Hanley. 2018. Rethinking "Democratic Backsliding" in Central and Eastern Europe - Looking beyond Hungary and Poland. *East European Politics* 34 (3):243–256. doi:10.1080/21599165.2018.1491401.

Čipin, I., A. Akrap, J. Knego, P. Međimurec, and K. Đurđević. 2014. *Strucna Podloga Za Izradu Strategije Prostornog Razvoja Republike Hrvatske: Demografski Scenariji I Migracije*. Zagreb: Sveuciliste u Zagrebu, Ekonomski fakultet, Katedra za demografiju. https://mgipu.gov.hr/UserDocsImages//dokumenti/Prostorno/StrategijaPR//Demografski_scenariji_i_migracije.pdf.

Cloke, P. J. 2006. Conceptualizing Rurality. In *Handbook of Rural Studies*, edited by P. Cloke, T. Marsden, and P. H. Mooney, 18–28. London: Sage Publications Ltd.

Copus, A., M. Shucksmith, T. Dax, and D. Meredith. 2011. Cohesion Policy for Rural Areas after 2013. A Rationale Derived from the EDORA Project (European Development Opportunities in Rural Areas) - ESPON 2013 Project 2013/1/2. *Studies in Agricultural Economics* 113:121–132. doi:10.7896/j.1113.

Cosier, J., E. Šabec, L. Verlič, A. Ponikvar, I. Jenko, K. Klemenčič, U. Gostonj, M. Kežar, M. Gamse, K. Uršič, J. Pavlovič, and I. Potočnik Slavič. 2014. Understanding Disparities in Slovenian Rural Areas: Various New Indicators. *Geoadria* 19 (2):149–164. doi:10.15291/geoadria.32.

Dax, T., and A. Copus. 2016. The Future of Rural Development. In *Research for AGRI Committee – CAP Reform Post-2020 – Challenges in Agriculture, Workshop Documentation*, 221–303. Brussels: Directorate-General for Internal Policies, Policy Department B, Agriculture and Rural Development, Structural and Cohesion Policies, European Parliament.

Dax, T., W. Strahl, J. Kirwan, and D. Maye. 2016. The Leader Programme 2007–2013: Enabling or Disabling Social Innovation and Neo-endogenous Development? Insights from Austria and Ireland. *European Urban and Regional Studies* 23 (1):56–68. doi:10.1177/0969776413490425.

EDORA [European Development Opportunities for Rural Areas]. 2011. Final Report, Parts A, B and C. https://www.espon.eu/sites/default/files/attachments/EDORA_Final_Report_Parts_A_and_B-maps_corrected_06-02-2012.pdf.

ESPON ESCAPE. 2019. European Shrinking Rural Areas: Challenges, Actions and Perspectives for Territorial Governance. Inception Report. https://www.espon.eu/sites/default/files/attachments/ESCAPE%20Inception%20Report_Final_13052019.pdf.

Ezcurra, R., P. Pascual, and M. Rapún. 2007. The Dynamics of Regional Disparities in Central and Eastern Europe during Transition. *European Planning Studies* 15 (10):1397–1421. doi:10.1080/09654310701550850.

Foa, R., and M. Howard. 2006. Use of Monte Carlo Simulation for the Public Sector: An Evidence-based Approach to Scenario Planning. *International Journal of Market Research* 48 (1):27–48. doi:10.1177/147078530604800103.

Franić, R., Ž. Jurišić, and R. Gelo. 2014. Food Production and Rural Development—Croatian Perspective within the European Context. *Agroeconomia Croatica* 4 (1):16–24.

Godet, M., and F. Roubelat. 1996. Creating the Future: The Use and Misuse of Scenarios. *Long Range Planning* 29 (2):164–171. doi:10.1016/0024-6301(96)00004-0.

Gordon, T. J., and J. Glenn. 2018. Interactive Scenarios. In *Innovative Research Methodologies in Management*, edited by L. Moutinho and M. Sokele, 31–61. Cham, U.K.: Palgrave Macmillan. doi:10.1007/978-3-319-64400-4_2.

Gorton, M., C. Hubbard, and L. Hubbard. 2009. The Folly of European Union Policy Transfer: Why the Common Agricultural Policy (CAP) Does Not Fit Central and Eastern Europe. *Regional Studies* 43 (10):1305–1317. doi:10.1080/00343400802508802.

Gullino, P., M. Devecchi, and F. Larcher. 2018. How Can Different Stakeholders Contribute to Rural Landscape Planning Policy? The Case Study of Pralormo Municipality (Italy). *Journal of Rural Studies* 57:99–109. doi:10.1016/j.jrurstud.2017.12.002.

Guštin, Š., and I. Potočnik Slavič. 2020. Conflicts as Catalysts for Change in Rural Areas. *Journal of Rural Studies* 78:211–222. doi:10.1016/j.jrurstud.2020.06.024.

Halfacree, K. 2006. Rural Space: Constructing a Three-fold Architecture. In *Handbook of Rural Studies*, edited by P. Cloke, T. Marsden, and P. H. Mooney, 44–62. London: Sage Publications Ltd.

Holman, I. P., C. Brown, V. Janes, and D. Sandars. 2017. Can We Be Certain about Future Land Use Change in Europe? A Multi-scenario, Integrated-assessment Analysis. *Agricultural Systems* 151:126–135. doi:10.1016/j.agsy.2016.12.001.

Kindu, M., T. Schneider, M. Döllerer, D. Teketay, and T. Knoke. 2018. Scenario Modelling of Land Use/Land Cover Changes in Munessa-Shashemene Landscape of the Ethiopian Highlands. *Science of the Total Environment* 622–623:534–546. doi:10.1016/j.scitotenv.2017.11.338.

Kisiała, W., and K. Suszyńska. 2017. Economic Growth and Disparities: An Empirical Analysis for the Central and Eastern European Countries Equilibrium. *Quarterly Journal of Economics and Economic Policy* 12 (4):613–631. doi:10.24136/eq.v12i4.32.

Kovács, K. 2010. Social and Administrative Crises Interlocking: The Misery of Rural Peripheries in Hungary. *Eastern European Countryside* 16:89–113. doi:10.2478/v10130-010-0005-5.

Loewen, B., and S. Schulz. 2019. Questioning the Convergence of Cohesion and Innovation Policies in Central and Eastern Europe. In *Regional and Local Development in Times of Polarisation. New Geographies of Europe*, edited by T. Lang and F. Go⊠rmar, 121–148. Singapore: Palgrave Macmillan. doi:10.1007/978-981-13-1190-1_6.

Lowe, P., and N. Ward. 2009. England's Rural Futures: A Socio-geographical Approach to Scenarios Analysis. *Regional Studies* 43 (10):1319–1332. doi:10.1080/00343400903365169.

Lukić, A., and O. Obad. 2016. New Actors in Rural Development—The LEADER Approach and Projectification in Rural Croatia. *Sociologija I Prostor* 54 (204–1):71–90. doi:10.5673/sip.54.1.4.

Macours, K., and J. F. M. Swinnen. 2008. Rural–Urban Poverty Differences in Transition Countries. *World Development* 36 (11):2170–2187. doi:10.1016/j.worlddev.2007.11.003.

Marquardt, D., S. Wegener, and J. Möllers. 2010. Does the EU LEADER Instrument Support Endogenous Development and New Modes of Governance in Romania? Experiences from Elaborating an MCDA Based Regional Development Concept. *International Journal of Rural Management* 6 (2): 193–241. doi:10.1177/097300521200600202.

OECD. 2006. *OECD Rural Policy Reviews: The New Rural Paradigm, Policies and Governance.* Paris: OECD Publishing.

Páthy, Á. 2017. Types of Development Paths and the Hierarchy of the Regional Centres of Central and Eastern Europe. *Regional Statistics* 7 (2):124–147. doi:10.15196/RS070202.

Perpar, A. 2014. Kljucni Dejavniki Razvojne Uspešnosti Podeželskih Obmocij V Sloveniji [PhD diss.], Univerza v Ljubjani, Biotehniška fakulteta.

Pociūtė-Sereikienė, G., and E. Kriaučiūnas. 2018. The Development of Rural Peripheral Areas in Lithuania: The Challenges of Socio-Spatial Transition. *European Countryside* 10 (3):498–515. doi:10.2478/euco-2018-0028.

Priess, J. A., J. Hauck, R. Haines-Young, R. Alkemade, M. Mandryk, C. Veerkamp, B. Gyorgyi, R. Dunford, P. Berry, P. Harrison, J. Dick, H. Keune, M. Kok, L. Kopperoinen, T. Lazarova, J. Maes, G. Pataki, E. Preda, C. Schleyer, C. Görg, A. Vadineanu, and G. Zulian. 2018. New EU-scale Environmental Scenarios until 2050 - Scenario Process and Initial Scenario Applications. *Ecosystem Services* 29 C:42–551. doi:10.1016/j.ecoser.2017.08.006.

Radeljak Kaufmann, P. 2016. Metoda Scenarija U Istraživanju I Planiranju prostora/Scenario Method in Spatial Research and Planning. *Hrvatski Geografski Glasnik* 78 (1):45–71. doi:10.21861/HGG.2016.78.01.03.

Rawluk, A., and A. Curtis. 2017. "A Mirror and a Lamp": The Role of Power in the Rural Landscape Trajectory of the Ovens Region of Australia. *Society & Natural Resources* 30 (8):949–963. doi:10.1080/08941920.2016.1264651.

Ray, C. 1999. Towards a Meta-Framework of Endogenous Development: Repertoires, Paths, Democracy and Rights. *Sociologia Ruralis* 39 (4):521–536. doi:10.1111/1467-9523.00122.

———. 2006. Neo-endogenous Rural Development in the EU. In *Handbook of Rural Studies*, edited by P. Cloke, T. Marsden, and P. H. Mooney, 278–291. London: Sage Publications Ltd.

Rienks, W. A. edited by 2008. *The Future of Rural Europe: An Anthology Based on the Results of the Eururalis 2.0 Scenario Study*. Wageningen: Wageningen University Research and Netherlands Environmental Assessment Agency.

Ringland, G. 2006. *Scenario Planning: Managing for the Future*. Chichester, U.K.: John Wiley & Sons, Ltd.

Rizov, M. 2006. Rural Development Perspectives in Enlarging Europe: The Implications of CAP Reforms and Agricultural Transition in Accession Countries. *European Planning Studies* 14 (2):219–238. doi:10.1080/09654310500418101.

Schoemaker, P. J. H. 1993. Multiple Scenario Development: Its Conceptual and Behavioral Foundation. *Strategic Management Journal* 14 (3):193–213. doi:10.1002/smj.4250140304.

Shearer, A. W. 2005. Approaching Scenario-based Studies: Three Perceptions about the Future and Considerations for Landscape Planning. *Environment and Planning. B, Planning & Design* 32 (1):67–87. doi:10.1068/b3116.

Sokol, M. 2001. Central and Eastern Europe a Decade after the Fall of State-Socialism: Regional Dimensions of Transition Processes. *Regional Studies* 35 (7):645–655. doi:10.1080/00343400120075911.

Soliva, R., K. Rønningen, I. Bella, P. Bezak, T. Cooper, B. E. Flø, P. Marty, and C. Potter. 2008. Envisioning Upland Futures: Stakeholder Responses to Scenarios for Europe's Mountain Landscapes. *Journal of Rural Studies* 24 (1):56–71. doi:10.1016/j.jrurstud.2007.04.001.

Spoor, M., L. Tasciotti, and M. Peleah. 2014. Quality of Life and Social Exclusion in Rural Southern, Central and Eastern Europe and the CIS. *Post-Communist Economies* 26 (2):201–219. doi:10.1080/14631377.2014.904107.

Swain, N. 2016. Eastern European Rurality in a Neo-Liberal, European Union World. *Sociologia Ruralis* 56 (4):574–596. doi:10.1111/soru.12131.

Trammell, E. J., J. S. Thomas, D. Mouat, Q. Korbulic, and S. Bassett. 2018. Developing Alternative Land-use Scenarios to Facilitate Natural Resource Management across Jurisdictional Boundaries. *Journal of Environmental Planning and Management* 61 (1):64–85. doi:10.1080/09640568.2017.1289901.

Van der Ploeg, J. D., and T. Marsden, edited by 2008. *Unfolding Webs: The Dynamics of Regional Rural Development*. Assen, Netherlands: Van Gorcum.

Van der Ploeg, J. D., H. Renting, G. Brunori, K. Knickel, J. Mannion, T. Marsden, K. de Roest, E. Sevilla-Guzman, and F. Ventura. 2000. Rural Development: From Practices and Policies Towards Theory. *Sociologia Ruralis* 40 (4):391–408. doi:10.1111/1467-9523.00156.

Varum, C. A., and C. Melo. 2010. Directions in Scenario Planning Literature—A Review of the past Decades. *Futures* 42 (4):355–369. doi:10.1016/j.futures.2009.11.021.

Woods, M. 2011. *Rural.* Abingdon, U.K.: Routledge.

Woods, M., B. Nienaber, and J. McDonagh. 2015. Globalization Processes and the Restructuring of Europe's Rural Regions. In *Globalization and Europe's Rural Regions*, edited by J. McDonagh, B. Nienaber, and M. Woods, 199–221. Farnham, U.K.: Ashgate.

Zagaria, C., C. J. E. Schulp, T. Kizos, D. Gounaridis, and P. H. Verburg. 2017. Cultural Landscapes and Behavioral Transformations: An Agent-based Model for the Simulation and Discussion of Alternative Landscape Futures in East Lesvos, Greece. *Land Use Policy* 65:26–44. doi:10.1016/j.landusepol.2017.03.022.

DOES LOCAL RURAL HERITAGE STILL MATTER IN A GLOBAL URBAN WORLD?

SERGE SCHMITZ and LAURIANO PEPE

ABSTRACT. Since 1976, the NGO "Qualité Village Wallonie" has worked to safeguard local rural heritage. Yet, over the years, the staff working at the NGO have noticed important changes in rural communities, including other perspectives on the local heritage. This may be linked with rapid changes in demography or changes of interests with globalization. This paper documents the apparent loss of sense of local heritage among inhabitants of the Belgian countryside and, in doing so, reflects on the sense of local places in an era of globalization. Based on interviews with the staff of the NGO and on an e-survey, the research presents a number of explanations for the findings. The results show a shift of attention away from religious and farming heritage toward places that make the village unique. The notion of heritage is evolving from something of the past that requires protection to things, sometimes intangible, that could be useful for the present and future generations. People claim that local heritage matters for the bonding and well-being of rural people.

Why should we take care of local heritage? While the trend in the scientific literature increasingly focuses on the tourist potential of heritage, we should acknowledge that most local heritage presents poor marketability for tourism development. However, at the same time, local heritage seems to be important for identity building. Tangible things help to construct personal, place, and community identity (Di Méo 1994; Harvey 2001). The stewardship of local heritage contributes to the social health of communities and empowers social capital and local identity (Stephens and Tiwari 2015). In globalized world marked by the homogenization of culture, local heritage is identified as a valuable resource to maintain and develop both the local economy and community (Brennan et al. 2009; Sacco et al. 2013). The bonding of the community is praised as a valuable objective, especially when it is associated with a sense of belonging, because it facilitates exchanges, social capital, and resilience. Local heritage can contribute to sharing a sense of place and finally to a sense of belonging to a community. It seems particularly relevant in rural areas faced with suburbanization and agriculture modernization.

Yet, we are reluctant to use the concept of community heritage in Belgium due to the fragmentation of the rural society where several practices and representations of "rural" coexist (Schmitz 2000). The model developed by Keith Halfacree (2006) describing the great diversity of motivations, representations, and practices in rural areas may be used to encompass this diversity. In practice, maintaining local heritage could be seen as a burden that increasingly fewer numbers of inhabitants want to support.

If the concern with heritage has been largely documented in the literature, we did not find a paper examining the end of the sense of heritage. This paper focuses on the heritage of local value, which receives less attention from official authorities, but is recognized by the local society as interesting and worth preserving for values that may differ from the authorized heritage discourse (Smith 2006; Mydland and Grahn 2012). In fact, any component of a landscape can become heritage (Di Méo 2007; Schmitz and Bruckmann 2020). Increasingly, the concept of heritage is being expanded from material and architectural elements to natural and intangible elements (ICOMOS 1999). Yet, in the process of scrutinizing the issue, it is important to separate the material matter that could be recognized as heritage from people's perceptions that shape, neglect, or forget this potential heritage. Heritage is not the object in itself, but the relationship with the object (Poulot 1993). Indeed, as André Gob (2017) argues, heritage value should be seen as a present value independent of the past values.

The Research Context

During the second part of the twentieth century, rural space in Belgium, a small country (30,500 km2 of territory and 11.5 million inhabitants) in the European Union with a predominantly urban population (91.5 percent) (OECD 2018), experienced a major decrease in the number of farmers and accommodated urban sprawl. Large sections of the population commute to cities for work, shopping, and recreation. At the same time, the rural economy experienced a diversification based on services, the primary sector, but also industrialization (Van Hecke et al. 2010). Concerning the demographic situation, rural space experienced, on one hand, a loss of active people and the aging of the population and, on the other hand, arrivals of commuters that led to a decline in the native population's share of the total people. In the more attractive landscapes around major cities, a residential economy prevails resulting in the consumption and the commodification of the rural space (Woods 2011).

In Wallonia (3.6 million inhabitants), the French speaking part of Belgium, the NGO "Qualité Village Wallonie" (QVW) has worked on safeguarding and highlighting local rural heritage since 1976. The QVW helped to develop 2,500 projects in 850 villages during recent decades. The staff of QVW has noticed important changes in rural communities, including alternative perspectives on their heritage. This could be related to the various modifications that affect the countryside and its population with rapid demographic changes, including aging and decline of the native population and the arrival of commuters. An alternative hypothesis to explain these differing perspectives is related to changes in people's lifestyles, particularly due to globalization and the digital revolution. Factors such as the rejection of Catholicism, increased mobility—including frequent leisure travel—global culture, social media, and the increase of consumerism could all contribute to the loss of the sense of local heritage in Belgian society.

Our research aims to document this apparent loss of sense of local heritage among inhabitants of the Belgian countryside and, in doing so, reflects on the sense of local places in an era of globalization. The research also aims to help QVW by analyzing the new demands of rural people. During the process of designing the methodology, the staff and the administrators of QVW took part in defining the research objectives and were the first informants and commentators of the results. In an increasingly urban and global world, what does the local rural heritage mean for both the native population and the newcomers? Can we explain the new perspectives on the local heritage by a loss of interest in these outdated elements? Or, do the new generations of inhabitants focus their attention on renewed sorts of tangible and intangible heritage? Our hypothesis is that in the past, heritage played a role that nowadays lacks relevance for people. Moreover, the paper questions the notions of centrality and locality in old rural settlements integrated in suburban areas. These questions go well beyond the scope of Wallonia. All over the world, there is an interest in maintaining some tangible or intangible elements of the community history to support the collective memory and to bind people together. The research conducted on the theme of heritage during recent decades shows the evolution of concerns. It started with pragmatic planning questions to introduce increasingly critical approaches (Veschambre 2007). Because heritage is culturally dependent (World Heritage Center 2008; Alberts and Hazen 2010), special attention will be given, in this research, to the literature in French. Based on interviews with the staff and administrators of QVW and on an e-survey of rural people, the research describes and suggests explanations for this apparent loss of sense of local heritage.

THE EMERGENCE OF A LOCAL HERITAGE

In Western Europe, the 1980s were marked by an interest in the notion of heritage, not only national heritage that makes kings and nations, but also broader expressions of heritage. After years of Americanization of European society, it has felt like a necessity to pay attention to heritage and to rediscover local and regional peculiarities. As a result of lifestyle changes, people's attention shifted toward noticing vernacular elements of their environment that they perhaps had not seen before. This leads to the emergence of new heritage, more "banal" and "quotidian," but more relevant for people due to stronger attachment and symbolic meanings (Jones 2017; Lekakis and Dragouni 2020).

The international literature on heritage has shown an increasing interest in local heritage defined, in essence, as having a value recognized by a village or a community (Carter and Bramley 2002), as well as in the population's participation in the production and preservation of heritage, considering heritage as a social process (for example, Waterton and Smith 2010; Mydland and Grahn 2012; Jones 2017). In particular, Siân Jones (2017) suggests looking beyond the intrinsic aspects of heritage, such as its esthetic value or its historic past, and to take into account more "intangible" values. This author refers to the concept of social values used in various ways, but that she defines as the result of "a collective attachment to place that embodies meanings and values that are important to a community or communities" (Jones 2017, 22). As a result, even if esthetic or historical interest may still be important in the production of heritage, the population will produce heritage for other reasons, such as to avert a threat or to satisfy a desire to belong to a group of individuals and to build an identity (Braaksma et al. 2016). Today, "cultural heritage is seen as an instrument for the development of social experiences, relations, exchanges and so forth" (Mydland and Grahn 2012, 583).

A GEOGRAPHICAL APPROACH

According to Brian Graham et al. (2000), geography and heritage intersect. First of all, heritage is a spatial phenomenon because it is embedded in space through its location, context, and spatial scale. Thus, the perception of heritage can vary according to the environment in which it is situated and the scale (local, regional, national, and international) at which we observe it. Secondly, cultural geography considers heritage as an identity component of a place and geographers are interested in studying specifically how the history of the place is expressed, particularly by the individual. Finally, heritage is considered as a resource of the territory, useful for the development and attractiveness of the territory. In this regard, heritage is a cultural and economic good to be commodified.

Looking at the French geographical literature dealing with heritage, Vincent Veschambre (2007) pointed to three periods. First, the geographers, who are more comfortable with the material world, examined particularly pragmatic issues of heritage relating to planning and local development: how to integrate heritage in the contemporary living space and how can heritage play a role in local development, especially through tourism development. Then, in the 1990s, an array of geographers influenced by new cultural geography paid increasing attention to the ontological value of heritage and brought out the links with personal and community identity. This work relates to the stance that heritage gives existence to the past, the territory, and the community (see Di Méo 1994). Finally, a critical current scrutinizes the political dimension of all heritagization (for example, Graham et al. 2000; Carter et al. 2019). Scholars paid particular attention to the following questions: Who makes the selection of heritage goods? For which purposes? What is the process? François et al. (2006) describe the heritagization stages, from the selection of goods and the justification to the conservation, the exhibition, and valorization. In each stage, the object changes in status or function or in both characteristics (Babelon and Chastel 1994). When heritagization becomes an official process, the agenda clearly defends the interest and priorities of the dominant class (Del Pozo and González 2012). Does such a relation exist when examining local heritage? Who could be the dominants and how do they exploit local heritage?

It is possible to summarize the approaches to heritage in a triangular diagram where the apex represents three visions of heritage: attachment, resource, and charge (Figure 1). The "attachment" dimension reveals people's sense of place. Are local people attached to the heritage element? Do they give their allegiance to it? Are they ready to sacrifice time and money? (Shamai 1991; Waterton 2005; Schmitz 2012) The "resource" dimension assesses the use, and sometimes the abuse, of heritage to contribute to extra incomes or a political agenda. Heritage is seen as an asset to economically develop the area or to consolidate an ideology. This can lead to commodification and the predation of the resource by a group of people (Landel and Senil 2009). The third dimension, "charge," less underlined by the literature (Caust and Vecco 2017), measures the feeling of burden. If people pay attention and maintain heritage, it is more experienced as a duty, a responsibility regarding the past and future generations. It may also be seen as an obstacle to new developments (Graham et al. 2005; Palazzo and Pugliano 2015). Depending on people's view and way of living, the sense of heritage for an object may move through this triangular space, giving more or less weight to each of the dimensions.

HERITAGE IN A RURAL CONTEXT

While the rural spaces evolve and the social value of heritage leads to the recognition of heritage as an evolutionary process (Jones 2017), the current interest in rural local heritage is frequently characterized by a lack of interest by the official authorities (Mydland and Grahn 2012) or by a weak recognition of the role of this heritage among a large part of the population (Hodges and Watson 2000). Rural heritage is often "described as carrying historic and/or scientific values and conceptualized as residing in physical elements and historic structures that were created by its surrounding communities and their activities on the landscape, including farming, animal husbandry or fishing" (Lekakis and Dragouni 2020, 86). However, rural

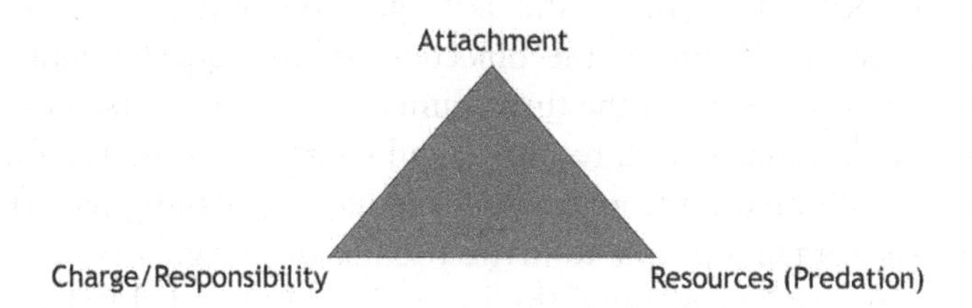

FIG. 1—The three dimensions of the sense of heritage (Schmitz and Pepe 2021).

heritage has often been considered as a "fixed" resource conserved and enhanced from the perspective of tourism development (González 2017) and from the commodification of the "countryside ideal" (Mitchell 1998; Kneafsey 2001; Woods 2011). More recently, Stelios Lekakis and Mina Dragouni (2020) support the idea that in order to better understand the construction of rural heritage, it is necessary to know the context in which it is integrated, and to consider this heritage "as a unified and intertwined system, highlighting the position of all nodes, such as material and intangible aspects, local and distant communities, social relations, and production processes . . . " (85).

Methods

This research aims to support the reflection of the NGO "Qualité Village Wallonie" (QVW). After more than 40 years of projects all over Wallonia, QVW questions both their actions and also the evolution of the sense of heritage (the current utility of local heritage and its future prospects) in suburban countryside marked by global culture. In order to support QVW in their questioning, we first conducted semistructured interviews and workshops with the staff and the administrators of QVW. Seven interviews were recorded and transcribed for analysis by both authors. Most of the people interviewed have several decades of experience in the domain and can trace people's interaction with local heritage over that time.

The insights gained from the interviews are used in conjunction with the results of an e-survey filled in by 1091 people from March to June 2020. A web page with a URL has been created via the online tool "framaforms" to host the survey. We restricted the number of entries per IP address to one. The questionnaire is short and playful to avoid participants abandoning it during the process. It allows people to reflect on their surroundings and their involvements in the local community. We restricted the identity questions to the name of the locality, the age, the gender, the period of living in the locality, the ownership of the dwelling, and the past contacts with QVW. The questionnaire invited participants to list emblematic places and four heritage elements of their municipality, and included 18 items rated on a Likert scale about the richness of the municipality from a heritage point of view, their personal involvement in heritage, the reasons for a moderate personal involvement, and their views on the contribution of heritage to the local life. Finally, the last question asked was, "Who should maintain local heritage?" One of the objectives of this questionnaire is thus to position individuals in regard to the three dimensions of the sense of heritage that have been presented (attachment, resource, and charge). We shared the invitation to fill out the questionnaire through social media and mailing lists from several organizations, encouraging people to invite friends and neighbors to complete the survey. The e-mail briefly explained the research and included links to the questionnaire, and both the university research unit and QVW websites. We regularly analyzed the characteristics of the respondents and informed the partners about gaps in our sample to relaunch invitations to underrepresented areas or ages.

Socio-demographic information on the survey respondents is presented in Table 1. Admittedly, the sample is not fully representative of the Belgian population living in the countryside. People who have answered are clearly more interested in heritage than the average Belgian. However, the survey gives an image of the situation and its main objective is to extrapolate the findings issued from the interviews to a larger group. The sample is reasonably balanced regarding gender, with a slight bias toward men (52.4 percent). With a mean age of 55 years, the sample is older than the Belgian population. Effectively, the respondents aged between 18 and 54 years were underrepresented, but the older (more than 54 years) were overrepresented. The survey gathers opinions of people from all Wallonia with an emphasis on the countryside. Respondents declare, 73.6 percent of the time,—that they live in the countryside and less than one-third think they are living in an urban or semiurban environment. Homeowners are also more represented (83.4 percent) than in the Belgian society (70.0 percent), but a part of this difference is explained by the emphasis on the countryside where the share of tenants is lower than in cities. Concerning the duration of residence, the majority of surveyed people have been living in their current place of residence for more than 20 years (62.9 percent). We also note that more than half of the sample have not been in contact with QVW in the past (62.3 percent).

TABLE 1—PROFILE SAMPLE (N = 1091)

ATTRIBUTE	CATEGORY	PERCENTAGE
Gender	Female	47.6
	Male	52.4
Age	15–24 years	5.4
	25–54 years	37.3
	55–64 years	28.0
	More than 64 years	29.3
Ownership of the housing	Owner	83.4
	Tenant	9.8
	Cohabitant	6.8
Declared living environment	Countryside	73.6
	City	18.5
	Periphery	7.9
Time of residence	Under than 1 to 5 years	11.3
	6–10 years	9.0
	11–20 years	16.8
	More than 20 years	62.9
Past contact with QVW	Yes	37.7
	No	62.3

Results

A CHANGING PERSPECTIVE ON HERITAGE

The interviews with the staff and the administrators of QVW pointed out that the interest in local rural heritage was more important in the 1980s and 1990s than in 2020. This differing interest could be explained by thinking of it as the life cycle of a product. People discovered their heritage in the 1980s and invested in knowing and protecting it; then as the novelty faded, they slowly turned their interest to other local and international concerns, such as climate change mitigation and the transition movements. The issue of the rural local heritage in Belgium was more relevant in the 1980s because it was a way to resist drastic changes of the countryside. It was seen as important to preserve some rural features to ensure rurality and the village identity. This was marked by an interest in places that lost their utility such as fountains, outdoor washhouses, chapels, churchyards, abandoned schools, and local train stops.

According to the survey, 85 percent of respondents acknowledged that their village has a rich heritage. Regarding the heritage elements listed, we noticed that castles and churches are the most cited elements, far more than other diverse elements including paths, landscape viewpoints, public squares, postindustrial sites, natural curiosities, war memorials, gastronomic specialties, folkloric events, and such. The vernacular heritage, which were the core business of QVW, is scarcer in this list. However, despite the existence of a certain interest in their heritage, the staff of QVW underline that people no longer know the function and the meanings of things around them. Indeed, local heritage has become part of a living environment where visual esthetic takes priority. For instance, a group of inhabitants wanted to restore a fountain to beautify a village, but they never considered restoring its water-supply function. Practically, this last proposal would cost more, would require some expertise and maintenance, and may cause public nuisances. For all these reasons, the fountain was transformed into a flower container. The symbol is maintained but the utility of delivering water does not represent enough interest. It is true that the new generations have no direct links to the old practices of peasantry; the local memory is fading. Several interviews underlined the prevalence of beauty and the feeling created by this beauty.

> I think that people are touched by the beauty, like to conserve traces from the past, the history. (Anne, 20 years with QVW)

> Heritage is often a sensation; it is something that is inside. Heritage is: What a beautiful thing! (Alain, 26 years with QVW)

A broader perspective of local heritage should require raising awareness for people.

> Heritage is a rather vague concept that must be explained through concrete examples. People come to a first meeting and say: I am invited to a meeting on the heritage of my village and I have thought carefully. I do not see it around me. (Anne, 20 years with QVW)

CONTRACTUAL INVOLVEMENT IN LOCAL HERITAGE

Sixty-one percent of people declare they are committed to the preservation of local heritage and to its enhancement. And concerning this commitment, the staff and administrators provided insight into a new sense of time and community engagement. Inhabitants seem very busy and when they accept to volunteer, they want to have a clear contract about their engagement. There is also a reluctance to commit oneself for a long period and, sometimes, they would like to be paid for their investments in the community. Therefore, fellow citizens are more selective and want to see fast results. This may be problematic because working on the restoration of heritage often requires time to study and reflect on the best solutions, but also to collect funds from both private and public actors. As with many places in the world, the Belgian countryside has entered a world of instantaneousness. People await a response to an e-notification in the next seconds, but forget to preserve a chapel or a stone wall constructed centuries before.

Instead of the formation of small groups of citizens who like to highlight the local heritage, QVW increasingly intervenes to support people who are attached to a precise place. They collaborate for a precise target; then, the group disbands. The staff and administrators also noticed an overspecialization of volunteers who increasingly collaborate within their professional competencies, avoiding discovering and being responsible for other tasks. The results of the survey support these views. Fifty-three percent of the people surveyed agree that their lack of time explains their relatively weak involvement in the safeguarding of local heritage (Figure 2). It is also interesting to note the diversity of responses concerning the lack of know-how. If most people have an interest in local heritage, 48 percent do not know how they can contribute to its enhancement. It seems that there are still some missions for QVW and similar organizations to support citizens to take care of their local heritage.

The results of the survey support the premise that the degree of investment seems to vary with age and with length of residence (Figure 3). Younger generations and newcomers might have less time to commit or less interest. On the contrary, older generations seem to be more concerned about their local heritage. This is probably a way to both recall memories and transmit a part of their identity to the next generation.

NEW WAYS OF DWELLING

The staff and administrators of the NGO also noted that the ways of dwelling have changed drastically and could explain the changes of interest for both the

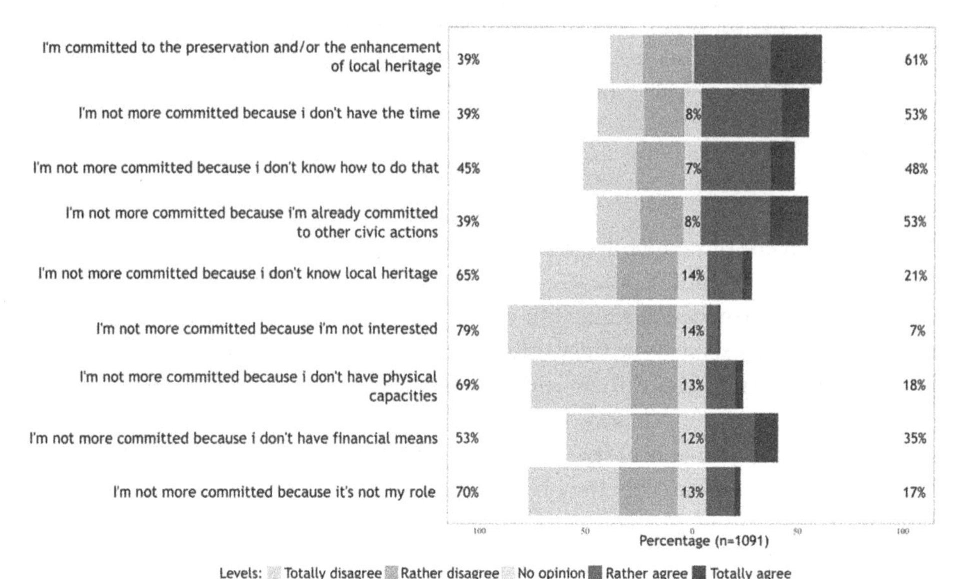

FIG. 2—Respondents' levels of commitment in local heritage.

local community and the heritage. Most new inhabitants are located on the periphery of the village, while older or poorer people inhabit the historic center. The houses in the historic center are often smaller, without a large garden, and require an extensive renovation to correspond to contemporary standards. So, the young generation from the village and newcomers inhabit peripheral districts. They are less and less confronted with historical heritage.

> In their neighborhood, there is no heritage because it is brand new. You can't make someone aware of something they don't have around them. (Alain, 26 years with QVW)

> Before, there was an interest because we lived in the village and we knew but now it's so distended, 2000 people, you don't know anymore, there is no longer this sense of local heritage in the living environment. (Isabelle, 32 years with QVW)

Considering social life, the staff and administrators underlined the impact of social media, which restrain the interactions with neighbors and consume an increasing share of the spare time.

> People have everything at their fingertips via internet, social networks. Even for contacts, they no longer need to see each other. It's so easy with the different technological means. Before, when they lived in a closed society, they all knew each other, they had activities within a tight geographical area. ... Heritage is also no longer necessarily one of the points of interest, they are no longer obliged to have a local hobby, so heritage is no longer a compulsory step to know his village. (Marie, 15 years with QVW)

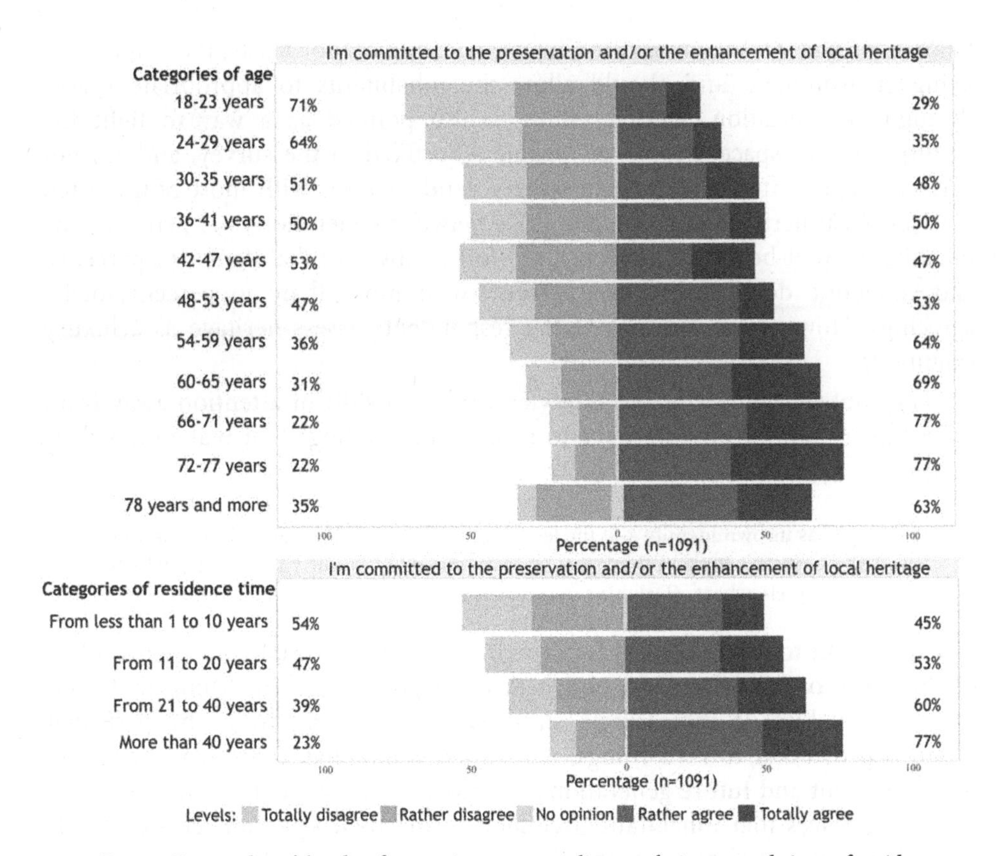

FIG. 3—Respondents' levels of commitment according to their age and time of residence.

Increasing mobility is another factor that could be used to explain the decreasing interest in the village. In many cases villages do not offer facilities such as shops, bars, and school, thus people spend their time elsewhere. Moreover, the decline of two practices reinforce this trend of disappearing interpersonal communication within the village community: the practice of Catholicism and of gardening. Nowadays, people choose to engage with people through new information and communication technologies rather than their neighbors. In the same way, they can also develop a heritage that does not need to be locally embedded via soap operas, world sport events, social media, and the like.

THE SOCIAL AND IDENTITY ROLE OF HERITAGE

Local historic heritage presents two weak points. Heritage could be described as an outdated issue, increasingly disconnected from contemporary life, and is frequently situated in the old village center, which has lost most of its facilities. Nonetheless, the staff of QVW pointed out that local heritage still provides some services: it contributes to the identity of the village, it allows the community to bond, especially on the occasion of inaugural ceremony or feasts, it enhances the living environment, and should allow the inhabitants to appropriate spaces through heritagization. Heritagization is also pointed as a way to fight the acculturation of space. Moreover, people who filled in the survey, and are not perfectly representative of Belgian society, tend to agree with most of the listed services: local heritage supports the links between generations (95 percent), the inhabitants' well-being (94 percent), the links between inhabitants (91 percent), and economic development (88 percent, with more than 40 percent totally agreeing). However, 18 percent of the respondents assess heritage as a luxury (Figure 4).

The content analysis of the interviews shows a shift of attention away from the religious and farming heritage to places and buildings that make the village unique.

> Each village has its own identity and the feeling of belonging to a village, to a community, it is also important for people to think that we have this thing in our village and maybe there is no such thing elsewhere. (Delphine, 16 years with QVW)

It is important to think of what is occurring in terms of more than simply a loss of the sense of local heritage, and instead to posit it as something that continuously evolves. It increasingly shifts from something from the past that requires protection toward things, sometimes intangible, that could be useful for the present and future generations. Moreover, these elements can be situated in remote places that inhabitants frequent during their open-air leisure. Beside a renewed interest for old paths that linked villages, local heritage gains in

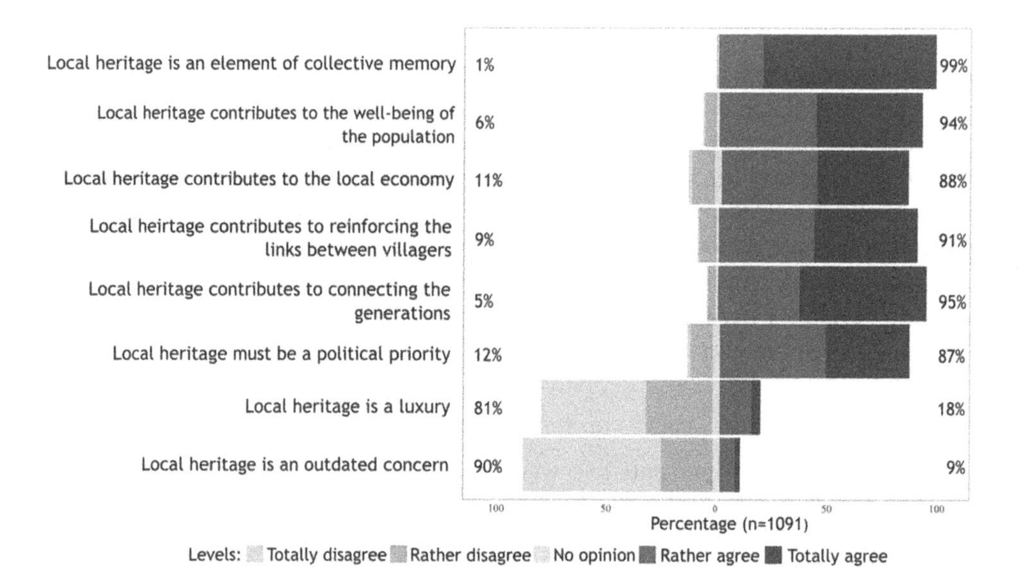

FIG. 4—Services delivered by local heritage according to survey respondents.

importance when it helps distinguish one village from the others. This trend is also visible in the reinvention of local folklore and patronal feasts, which give a pretext to organize parties. While losing their roots embedded in peasantry and Catholicism, by adopting the global and virtual culture and economy, people are looking to new elements that help them to be distinct.

> Heritage must speak to me and if possible, speaks to me about myself. (Marie, 15 years with QVW)

The idea that heritage is both a common good and an individual good, which gives rise to a social interaction, is an interesting one. The enhancement of a heritage (through its restoration or during folklore events) will be done by the will of a group of inhabitants. Still today, through this common will, local heritage allows people to participate in the collective life and to develop an identity. They are linked to the group that appropriates it, both from an emotional and symbolic point of view, and thus are able to distinguish this group of inhabitants, this neighborhood, this village from the others. Additionally, heritage can contribute to shape individual identity. An individual's identity is not limited to the relationships with others, but also to the relationships with the elements that constitute his or her environment (Proshansky et al. 1983).

> If they are not natives or linked to an old family from the village, some people will get involved in the preservation of the heritage of their new living environment. They will try to take a place because they have no place. When they get involved, it is to say: we are interested in our living environment, we do not come as invaders to remove everything. (Marie, 15 years with QVW)

Staff and administrators of the NGO like also to emphasis that acknowledging and taking care of heritage is also to transmit a message.

> I think people like it because they all have some information to bring and memories to tell they can convey. (Isabelle, 32 years with QVW)

> Our heritage, that's it, it conveys a message as well. (Delphine, 16 years with QVW)

WHO IS RESPONSIBLE?

While 87 percent of the respondents to the survey consider that local heritage should be placed higher on the political agenda, the question, "Who should be in charge of local heritage?" is a central question. In Belgium, this responsibility can be shared between the owner, the state (Wallonia), the municipality, or the inhabitants. Because the regional state invests mainly in monumental heritage of regional and international interest, the municipality may be the appropriate level to handle local heritage, but it lacks the income required to cover these expenses. Citizen involvement is needed if they want to preserve local heritage. Indeed, they have the opportunity to symbolically appropriate some places by investing time, competencies, and money, and so to maintain, or create in the case of newcomers, personal links with the village and the community. Ninety-one percent of the people surveyed pointed to municipalities as the entity principally responsible, which was followed by the state (80 percent), owners (69 percent), citizens (68 percent), and actors of the tourism sector (60 percent).

Another concern pointed toward in the interviews was the crosscutting of heritage policy. While heritage could be restricted to historic monuments and be managed by one dedicated administration, it also concerns planning, economic development, tourism, social cohesion, people identity, culture, and education. This richness could become a nightmare when searching for support from the authorities.

Discussion

To reflect on the insights gathered during the interviews and the survey relating to the sense and caring for local heritage, it is important to go back to the facets of the meaning of places (Relph 1976). We suggest dividing the meaning of place into six components: materiality, location, functions, people, history and myths, and finally, personal memories (Schmitz 2017). Paying attention to these components helps us to understand the notion of an end of the sense of local rural heritage.

The materiality of the object contributes both to its recognition and to the difficulties to maintain it. Belgians tend to list, first, castles and other important buildings as elements of their local heritage, but they are not able to find resources to preserve them. They seem to accord less importance to small pieces of rural landscape that require more awareness raising and interpretation. The location is another concern, as we have seen the rural local heritage migrates with the installation of the new generation in peripheral districts. The centers of the village are increasingly meaningless for this new generation and have no more functional interest. The people who live there are old and do not share the same interest as young generations. The social value concerning the local heritage located in the village center is not shared by all the inhabitants and is disappearing with the aging of the ancients. Moreover, newcomers and young generations do not know the histories of the buildings and places and so are not aware of their richness. They also have less time due to the family and professional activities, which are often located in towns. Finally, and probably most importantly, they do not associate these places with significant events in their lives.

Newcomers are becoming the dominant class and increasingly shape local policy and the image of the countryside. In many cases, the sense of heritage moves beyond simple and singular representations of the past to encompass peculiar elements of the environment that do not have a long history. The definition of heritage is broadening and increasingly pays attention to previously marginalized forms of heritage, such as paths that are inherited from medieval times. These can be used by present inhabitants, thus including "practices that might construct heritage implicitly, such as outdoor recreation" (Braaksma et al. 2016, 75).

Was the interest in local heritage a passing fad that helped people to face a crisis of identity around the 1980s? Rural local heritage anchored in preindustrial times was very relevant during periods of crisis, in face of the urbanization of the countryside in trying to maintain rural identity. Heritage concern emerges from emergency due to the fear of losing important elements of identity (Gob 2017). However, the new generation, with their new lifestyle, pay less attention to this rurality and have no more direct contact with people who were peasants. This heritage of previous generations may be seen as well as an obstacle to the restructuring of the village. It became a burden. The new generation broadens their interests to other places and other themes, and seem less concerned by local objects arising from an unknown past. Alain Touraine (1973, 7) already pointed out 50 years ago: "To know its present, society looks less and less to its past, because the share of what is transmitted continues to decrease compared to the share of what is acquired." Does it mean that there is a real loss of sense of rural local heritage in Belgium because the value that it represented disappeared, because its potential for commodification is weak? Jean Pierre Babelon and André Chastel (1994) noticed that the sense of heritage is based sometimes on the traditional values, which are attached to it and explain it, but sometimes and increasingly in the name of a new feeling that bonds the community. Yet, the low interest in local heritage is also a sign of the fragmentation of the rural society that does not share common values and expectations for the local landscape (Rivera Escribano and Mormont 2007; Halfacree and Rivera 2012).

The Belgian countryside, as with other rural spaces around the world, has changed drastically in terms of demographics, economic function, land use, and in cultural points of view. If villages and rural landscapes incorporate landmarks from the past, which were seen as heritage by the previous generations, the use and even the meanings of numerous elements are being forgotten. These pieces of heritage do not represent a lot to present inhabitants and, so, lose their heritage values. Even if the people who participated in the survey state having an interest in local heritage, they do not know whose is responsible and suggest a broad sharing of responsibility. We noticed also a specialization of personal interest that is directed to a special element instead of the village, its inhabitants, and its surroundings as a whole. Thanks to the esthetic or peculiarities of some of these elements, individuals notice them and try to safeguard them. Heritage, and the negotiation regarding its use, the duty to protect it, and its links with society are in a process of continuous construction and deconstruction of meaning (Harvey 2001; MacDonald 2013; Carter et al. 2019). Looking at the interest in a local piece of heritage, it is important to pay more attention to the possible exploitation of it. If people do not see possible usages, it lacks interest. If they do not know its meanings, they are less equipped to like it. The Belgian countryside is becoming a decor reflecting an urban way of life where heritage should be both documented by historical research and reinterpreted considering new perspectives of the population.

The idea that, concerning common heritage, we would not be owner but responsible (Landel and Senil 2009) seems inappropriate outside the cenacles who list the world and national heritage. The debate between those who prescribe maintaining historical structures and those who accept to adapt structures to contemporary life (Tyler 1999; Alberts and Hazen 2010) should encompass the discussion about the sense that people give to their places. Heritagization is a process of rethinking history and ideological symbols, and allows people to feel the progress of the society (Ilchenko 2020). Indeed, the spirit of heritage cannot be transmitted; it must be discovered and grasped by each generation and be rooted in the present (Bortolotto 2011). It becomes a heritage if people commit themselves. Globalization of the economy, cultural change in society and people's identity, including in villages, is more or less connected to cities. New technologies contribute to reduce the anchoring of rural people to their territory. However, four out five respondents to the survey, probably tens of thousands of people from Wallonia, acknowledge that they should pay attention to their heritage. While the notion of heritage is changing all the time and the value of landmarks can be totally different from the primary intention, the construction of local heritage seems to matter for rural people and contribute to the bonding and the well-being of the population.

Acknowledgments

We thank the staff and administrators of QVW for their contribution to this research and Michaela Korodimou and Mary Cawley for language editing of the manuscript.

ORCID

Serge Schmitz http://orcid.org/0000-0002-3412-818X

Lauriano Pepe http://orcid.org/0000-0002-6243-1960

References

Alberts, H. C., and H. D. Hazen. 2010. Maintaining Authenticity and Integrity at Cultural World Heritage Sites. *Geographical Review* 100 (1):56–73. doi:10.1111/j.1931-0846.2010.00006.x.

Babelon, J. P., and A. Chastel. 1994. *La notion de patrimoine.* Paris: Liana Levi.

Bortolotto, C. 2011. Le trouble du patrimoine culturel immatériel. *Terrain* 26:21–42.

Braaksma, P. J., M. H. Jacobs, and A. N. Van der Zande. 2016. The Production of Local Landscape Heritage: A Case Study in the Netherlands. *Landscape Research* 41 (1):64–78. doi:10.1080/01426397.2015.1045465.

Brennan, M. A., C. G. Flint, and A. E. Luloff. 2008. Bringing Together Local Culture and Rural Development: Findings from Ireland, Pennsylvania and Alaska. *Sociologia Ruralis* 49 (1):97–112. doi:10.1111/j.1467-9523.2008.00471.x.

Carter, R. W., and R. Bramley. 2002. Defining Heritage Values and Significance for Improved Resource Management: An Application to Australian Tourism. *International Journal of Heritage Studies* 8 (3):175–199. doi:10.1080/13527250220000/18895.

Carter, T., D. C. Harvey, R. Jones, and I. J. M. Robertson. 2019. *Creating Heritage: Unrecognised Pasts and Rejected Futures.* London: Routledge.

Caust, J., and M. Vecco. 2017. Is UNESCO World Heritage Recognition a Blessing or Burden? Evidence from Developing Asian Countries. *Journal of Cultural Heritage* 27:1–9. doi:10.1016/j.culher.2017.02.004.

Del Pozo, P. B., and P. A. González. 2012. Industrial Heritage and Place Identity in Spain: From Monuments to Landscapes. *Geographical Review* 102 (4):446–464. doi:10.1111/j.1931-0846.2012.00169.x.

Di Méo, G. 1994. Patrimoine et territoire, une parenté conceptuelle. *Espaces et Sociétés* 4:15–34. doi:10.3917/esp.1994.78.0015.

———. 2007. Processus de patrimonialisation et construction des territoires. In *Colloque Patrimoine et industrie en Poitou-Charantes: Connaître pour valoriser*, 87-109. Poitier-Châtellerault, France: Geste éditions.

François, H., M. Hirczak, and N. Senil. 2006. Territoire et patrimoine: La co-construction d'une dynamique et de ses ressources. *Revue d'Économie Régionale & Urbaine* décembre 5:683–700. doi:10.3917/reru.065.0683.

Gob, A. 2017. Patrimoine et patrimonialisation. In *Patrimoine culturel immatériel*, edited by F. Lempereur, 15–22. Liège, Belgium: Presses Universitaire de Liège.

González, P. A. 2017. Heritage and Rural Gentrification in Spain: The Case of Santiago Millas. *International Journal of Heritage Studies* 23 (2):125–140. doi:10.1080/13527258.2016.1246468.

Graham, B., G. Ashworth, and J. Tunbridge. 2000. *A Geography of Heritage*. London: Routledge.

———. 2005. The Uses and Abuses of Heritage. In *Heritage, Museums and Galleries: An Introductory Reader*, edited by G. Corsane, 28–40. London: Routledge.

Halfacree, K. H. 2006. Rural Space: Constructing a Three-fold Architecture. In *The Handbook of Rural Studies*, edited by P. Cloke, T. Marsden, and P. Mooney, 44–62. London: SAGE.

Halfacree, K. H., and M. J. Rivera. 2012. Moving to the Countryside … and Staying: Lives beyond Representations. *Sociologia Ruralis* 52 (1):92–114. doi:10.1111/j.1467-9523.2011.00556.x.

Harvey, D. C. 2001. Heritage Pasts and Heritage Presents: Temporality, Meaning and the Scope of Heritage Studies. *International Journal of Heritage Studies* 7 (4):319–338. doi:10.1080/13581650120105534.

Hodges, A., and S. Watson. 2000. Community-based Heritage Management: A Case Study and Agenda for Research. *International Journal of Heritage Studies* 6 (3):231–243. doi:10.1080/13527250050148214.

ICOMOS [International Council on Monuments and Sites]. 1999. International Cultural Tourism Charter. Managing Tourism at Places of Heritage Significance. https://www.icomos.org/charters/tourism_e.pdf.

Ilchenko, M. 2020. Working with the Past, Re-discovering Cities of Central and Eastern Europe: Cultural Urbanism and New Representations of Modernist Urban Areas. *Eurasian Geography and Economics* 61 (6):763–793. doi:10.1080/15387216.2020.1785907.

Jones, S. 2017. Wrestling with the Social Value of Heritage: Problems, Dilemmas and Opportunities. *Journal of Community Archaeology & Heritage* 4 (1):21–37. doi:10.1080/20518196.2016.1193996.

Kneafsey, M. 2001. Rural Cultural Economy: Tourism and Social Relations. *Annals of Tourism Research* 28 (3):762–783. doi:10.1016/S0160-7383(00)00077-3.

Landel, P.-A., and N. Senil. 2009. Patrimoine et territoire, les nouvelles ressources du développement. *Développement durable et territoires* (12). doi:10.4000/developpementdurable.7563.

Lekakis, S., and M. Dragouni. 2020. Heritage in the Making: Rural Heritage and Its Mnemeiosis at Naxos Island, Greece. *Journal of Rural Studies* 77:84–92. doi:10.1016/j.jrurstud.2020.04.021.

MacDonald, S. 2013. *Memorylands, Heritage and Identity in Europe Today*. London: Routledge.

Mitchell, C. 1998. Entrepreneurialism, Commodification and Creative Destruction: A Model of Post-modern Community Development. *Journal of Rural Studies* 14 (3):273–286. doi:10.1016/S0743-0167(98)00013-8.

Mydland, L., and W. Grahn. 2012. Identifying Heritage Values in Local Communities. *International Journal of Heritage Studies* 18 (6):564–587. doi:10.1080/13527258.2011.619554.

OECD [Organisation for Economic Co-operation and Development]. 2018. OECD Regional Statistics (Database). http://dx.doi.org/10.1787/region-data-en.

Palazzo, A. L., and A. Pugliano. 2015. The Burden of History: Living Heritage and Everyday Life in Rome. In *Theory and Practice in Heritage and Sustainability*, edited by E. Auclair and G. Fairclough, 54–68. London: Routledge.

Poulot, D. 1993. Le sens du patrimoine: Hier et aujourd'hui (note critique). *Annales. Économies, sociétés, civilisations* 48 (6):1601–1613. doi:10.3406/ahess.1993.279233.

Proshansky, H. M., A. K. Fabian, and R. Kaminoff. 1983. Place-identity: Physical World Socialization of the Self. *Journal of Environmental Psychology* 3 (1):57–87. doi:10.1016/S0272-4944(83)80021-8.

Relph, E. 1976. *Place and Placelessness*. London: Pion.

Rivera Escribano, M. J., and M. Mormont. 2007. Neo-rurality and the Different Meanings of the Countryside. In *Les mondes ruraux à l'épreuve des sciences sociales*, edited by C. Bessière, E. Doidy, L. Jacquet, G. Laferté, and Y. Sencebé, 33–45. Paris: INRA.

Sacco, P. L., G. Ferilli, G. T. Blessi, and M. Nuccio. 2013. Culture as an Engine of Local Development Processes: System-Wide Cultural Districts I: Theory. *Growth and Change* 44 (4):555–570. doi:10.1111/grow.12020.

Schmitz, S. 2000. Modes D'habiter Et Sensibilités Territoriales Dans Les Campagnes Belges. In *Des campagnes vivantes: Un modèle pour l'Europe?* edited by N. Croix, 627–632. Rennes, France: Presses Universitaires de Rennes.

———. 2012. Un Besoin De Territoire À Soi: Quelques Clés Pour Un Aménagement Des Espaces Communs. *Belgeo* (1–2). doi:10.4000/belgeo.6627.

———. 2017. Inscription territoriale et esprit du lieu. In *Patrimoine culturel immatériel*, edited by F. Lempereur, 55–64. Liège, Belgium: Presses Universitaire de Liège.

Schmitz, S., and L. Bruckmann. 2020. The Quest for New Tools to Preserve Rural Heritage Landscapes. *Documents d'Analisi Geografica* 66 (22):445–463. doi:10.5565/rev/dag.593.

Shamai, S. 1991. Sense of Place: An Empirical Measurement. *Geoforum* 22 (3):347–358. doi:10.1016/0016-7185(91)90017-K.

Smith, L. 2006. *Uses of Heritage*. London: Routledge.

Stephens, J., and R. Tiwari. 2015. Symbolic Estates: Community Identity and Empowerment through Heritage. *International Journal of Heritage Studies* 21 (1):99–114. doi:10.1080/13527258.2014.914964.

Touraine, A. 1973. *Production de la Société*. Paris: Le Seuil.

Tyler, N. 1999. *Historic Preservation: An Introduction to Its History, Principles and Practice*. New York: W. W. Norton & Company.

Van Hecke, E., M. Antrop, S. Schmitz, M. Sevenant, and V. Van Eetvelde. 2010. *Atlas de Belgique: Paysages, Monde rural et Agriculture*. Ghent, Belgium: Academia Press.

Veschambre, V. 2007. Patrimoine: Un objet révélateur des évolutions de la géographie et de sa place dans les sciences sociales. *Annales de géographie* 4 (656):361–381. doi:10.3917/ag.656.0361.

Waterton, E. 2005. Whose Sense of Place? Reconciling Archaeological Perspectives with Community Values: Cultural Landscapes in England. *International Journal of Heritage Studies* 11 (4):309–325. doi:10.1080/13527250500235591.

Waterton, E., and L. Smith. 2010. The Recognition and Misrecognition of Community Heritage. *International Journal of Heritage Studies* 16 (1–2):4–15. doi:10.1080/13527250903441671.

Woods, M. 2011. *Rural*. London: Routledge.

World Heritage Center. 2008. Operational Guidelines for the Implementation of the World Heritage Convention. UNESCO: World Heritage Center. https://whc.unesco.org/en/guidelines/.

YOUNG PEOPLE'S VISIONS FOR LIFE IN THE COUNTRYSIDE IN LATIN AMERICA

A. CRISTINA DE LA VEGA-LEINERT, JULIA KIESLINGER, MARCELA JIMÉNEZ-MORENO and CORNELIA STEINHÄUSER

ABTRACT. Young people steer their life trajectories against the backdrop of profound transformations of rurality. They are considered protagonists in the present and future of rural areas; however, their experiences and perspectives on life in, and away from, the countryside have been scarcely acknowledged in academia. Considering conceptual thoughts on new mobilities, place-based belonging, and new ruralities, we argue that young people are a highly dynamic and continually reassessing group. In four case studies realized independently in Argentina, Bolivia, Ecuador, and Mexico, we closely involved young people with multiple participative methods to express their perceptions, concerns, and demands. We experiment on ways to integrate our insights for future comparative studies. Beyond the diversity in local contexts, we found convergences in young people's visions, which motivate them to stay or leave, with wider implications for the sustainability of rural places. "Living in between" thereby encapsulates multifaceted hybrid possible countrysides away from constraining dichotomic views.

In many Latin American countries, rural out-migration, especially of young people, is discussed as a constant (Cazzuffi and Fernández 2018). This is increasingly addressed as a problem or a threat to be contained (Fabregat et al. 2020), rather than a potential to be fostered. While there seems to be a consensus on the multidimensional challenges young people face in countrysides (Guiskin et al. 2019), much effort is directed at pro-moting young people as actors of sustainable transformations in their rural communities (OECD 2018; UN 2018), especially because their visions can have a decisive role for the future of the countryside. While the literature on the food and conservation nexus envisages different avenues for local rural populations (de la Vega-Leinert and Clausing 2016), the sustainability paths debated in international policy (within the Sustainable Development Goals, United Nations 2015) require an urgent departure from current models of land use. The starting point of our reflection is the premise that a transition toward sustainability implies attending inter- and intragenerational perspectives, and fostering a population committed to its countryside, but also a society committed to its youth, its needs, demands, and visions of future.

Recent research highlights the contradiction that lies in perceiving youth as a strategic social group for promoting change, while largely ignoring rural young people, their experiences of places and perspectives on their (future) lives (Schmuck 2019). This is even more so, since youth is systematically marginalized and facing substantial challenges in the realization of individual life projects in rural areas and beyond (Rodríguez 2011). Young people's perspectives were, for example, largely underrepresented during the recent conference on "Visions of Future(s) in the Americas"[1] held in Bonn, Germany, in July 2019. Our core motivation is, therefore, to contribute to give young people an arena to voice their demands and concerns regarding the countryside they want today and in future.

When focusing on young people we should consider the social, historical, and cultural construction of youth, which varies in time and space (see Feixa 1999). Young people steer their life trajectories against the backdrop of profound transformations of rurality in a context of globalization. Despite having higher education levels compared to previous generations (Krauskopf 2005), young people often face great obstacles in accessing land, financial services, and well-paid jobs. Further, they must navigate intergenerational tensions, which stem from the deconstruction of traditional rural identities and the emergence of new hybrid ones (Urteaga and García 2015).

This paper explores these fields of tension based on four case studies done independently in rural areas of north Argentina, east Bolivia, south Ecuador, and south Mexico (Figure. 1), which were performed with the close involvement of young people. Our common transversal objectives are first, to better understand how young people perceive past, present, and future countrysides,[2] and how these perspectives are interrelated with negotiations on leaving and staying. Secondly, we aim at contributing toward a more comprehensive, multifaceted view of young people in sustainable transformation processes affecting local ways of life, means of livelihoods, and food security.

In this context, the research questions guiding our work are: How do young people perceive life in rural places in relation to other places? What motivates young people to stay and/or leave? What are the implications of young people's migration for rural livelihoods? And finally, which crosscutting themes emerge from our case studies in relation to young people's visions of the countryside?

After introducing theoretical and methodological considerations, we present complementing results from our case studies, explore emerging commonalities and divergences, and discuss ways to integrate insights gained as a basis for future comparative studies.

(Im)mobilities, Youth, and Rurality

With respect to young people's migration, we embed our reflection within the new mobilities paradigm (see Sheller and Urry 2006). Migration processes are here considered as human mobilities and addressed in a continuum from mobility to fixity. Being mobile and staying at places are considered as mutually interrelated (see Bell and Osti 2010). Further, mobilities are not constrained to physical movements but also encompass social meanings (Massey et al. 1993),

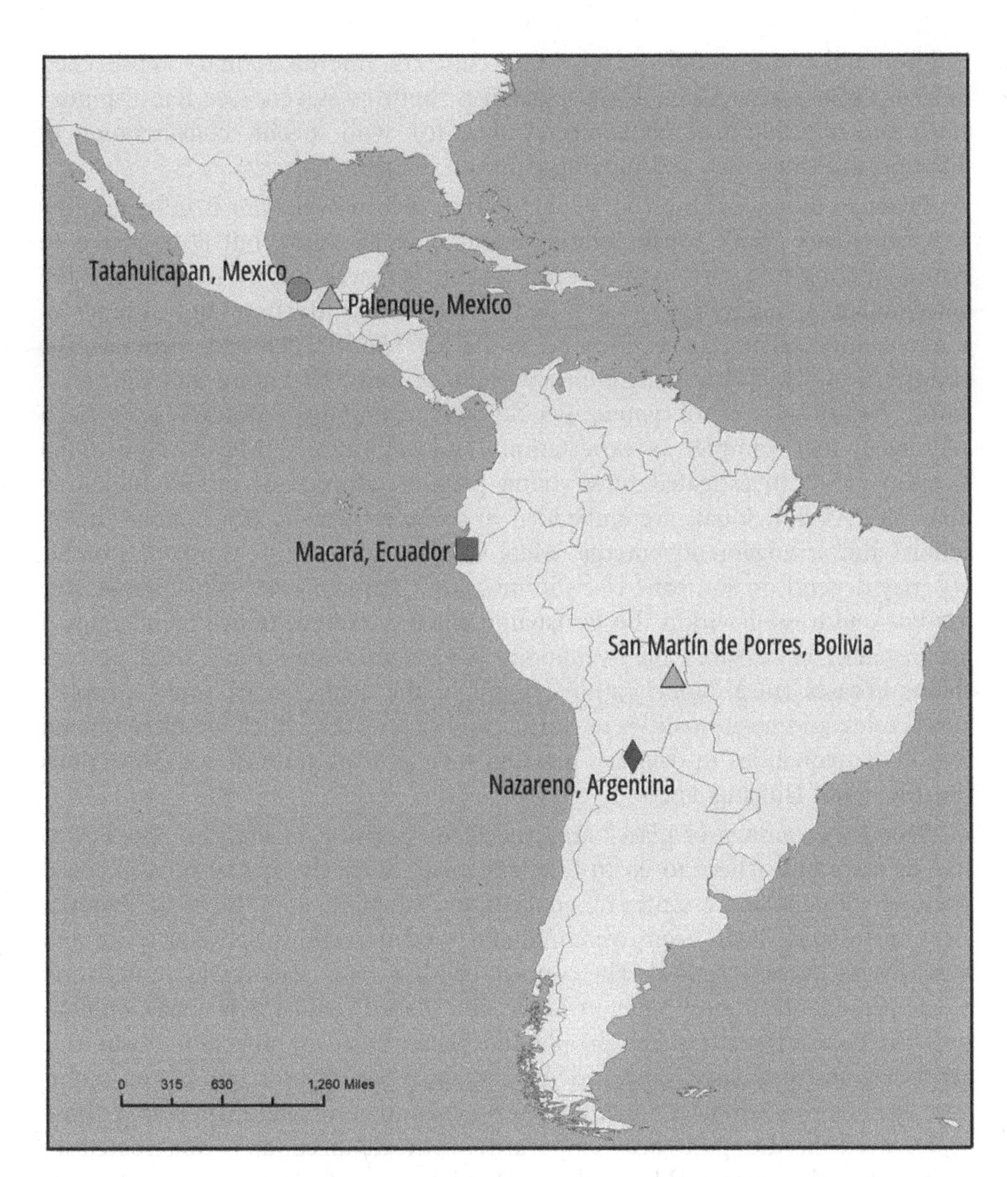

FIG. 1—Location map showing the four case studies in five places.

which are constantly negotiated via relational coconstructions of (life)times, places, and social interactions (Kieslinger et al. 2020).

Young people and their unique life experiences have often been overlooked in research, because of their assigned "positionality as 'children' in homes, in communities and in the nation" (Chakraborty and Thambiah 2018, 583). At the same time, scholars increasingly challenge idealized interpretations of childhood and youth, as fixed or bounded (Punch 2001; Farmer 2017). However, there is a growing area of study dedicated to better understanding the experiences of children and young people in migration contexts. Interdisciplinary approaches, indeed, explore young people's perspectives, both as stayers (see Rae-Espinoza 2016) and migrants (see Ní Laoire et al. 2010), with specific consideration of different age categories and moving patterns (Huijsmans 2017).

From an individual life-course perspective, moving behavior is influenced by past experiences with (im)mobilities (Coulter et al. 2016), but also related to future options and visions (see Carling and Schewel 2017). Accordingly, the determinants of migration, as well as the likelihood, direction, and distance of a move might change over lifetime, while age-specific demands generate the objectives to which individuals orient their actions (Ní Laoire and Stockdale 2016). Future visions of young people in (im)mobility contexts have been addressed, for example, as expectations and aspirations (Meyer 2017). Gina Crivello (2015) investigated how young people's aspirations unfold in family networks. As individuals are embedded in social structures, (im)mobility interactions and arrangements emerge (Mata-Codesal 2015), while someone's mobility may depend on someone else's immobility (Bauman 1998). This perspective has been addressed within the transnationalism paradigm (Glick Schiller 1997) for instance, in research on livelihoods (Geiger and Steinbrink 2012) or care chains (Yeates 2012). Some authors highlight the centrality of young people's views, roles, and positionalities in (im)mobility networks, which are often related to power inequalities in decision making, societal expectations, and aspirations (Bushin 2009; Huijsmans 2017).

People's meanings of places are crucial for decision making on whether to stay or leave and where to go to (Samers 2010). Since the spatial turn in social sciences (Soja 1989), essentialist conceptions of place as a physical material location in space have been overcome and the interactions between space and society have become central. The concept of place-based belonging, understood as the personal feeling of being at home in a place (Yuval-Davis 2006), emphasizes the connections between people and places based on affectivity, cognitive processes, and practices (Low and Altman 1992; Steinhäuser 2020b). As mobilities generate experiences in other places, meanings of place, including those experienced in the past, are temporarily contextualized and constructed in relation to other places (Murdoch 2005). Meanings of places of young people have been highlighted by Stuart C. Aitken as they "play a large part in constructing and constraining dreams and practices" (2001, 20). They affect young people's aspirations of future living while out-migration may be seen as constitutive for desirable identities (for example, for self-development) and result in ambivalent relationships with the place of origin and destination (for rural-urban migration context, see Dalsgaard Pedersen 2018).

However, (im)mobilities go along with social and environmental transformation processes of space-society interactions. In rural landscapes with subsistence agriculture, out-migration as well as changing lifestyles may lead to fragmentation of the familiar and social cohesion that is required to maintain the common resources and livelihoods (Ostrom 2000). This can be abandonment of fields and infrastructures (like irrigation channels), loss of agro-biodiversity, and local expertise, thereby reducing the complexity, resilience, and sustainability of the landscapes.

In this sense, (im)mobilities may also play a strategic role in livelihoods and food security,[3] particularly for households and families who depend on smallholder agriculture. Migration, locally an important social and economic livelihood alternative, relates to the generation and mobilization of economic capital, such as remittances, but also of human and social capital, which may have an impact on land uses, farming productivity, and subsistence strategies (Hecht et al. 2015). In this respect, Latin American scholars have coined the concept "new rurality" to emphasize how rural areas are being re-configured via the diversification of livelihoods, forms of (im)mobilities and changing sociocultural practice (Kay 2008; Appendini and Torres-Mazuera 2008). In this context, the expansion of formal schooling and its intricate linkages to (im)mobilities have been recognized as particularly important factors in shaping young people's life expectations and trajectories (Urteaga and García 2015). Both migration experiences and formal schooling—at place and abroad—may open alternative, nontraditional spaces for young people to explore and (re)build their identities, aspirations, and gender relationships (Urteaga and García 2015). Further, while social meanings and narratives of "the rural" are considered as constitutive for (im)mobilities, these, in turn, influence people's experience of rural living and conceptions of place and identity (Milbourne and Kitchen 2014). These relationships need to be explored in detail for specific social groups and in a broad range of rural places (Milbourne and Kitchen 2014). By presenting young people's perspectives from different Latin American contexts, this paper is a step in this direction.

Methodology

We share common principles that determine our research practices regarding researcher's positionality, research procedures, and ethical concerns. These include the construction of reality as a basis, understanding as an epistemological principle, everyday practices and knowledge as a focus, the contextuality of social experience, and researchers' reflexivity (Flick 2018).

A common characteristic of our case study regions (Figure 1) is that local livelihoods are primarily based on small-scale agriculture, livestock, and agroforestry for subsistence and sale at local markets. Tatahuicapan in south Mexico is a transition zone between volcanoes of Sierra de Santa Marta and the coastal lowlands of the Gulf of Mexico, with tropical rainforest and deciduous forests (Jiménez-Moreno Forthcoming). Palenque (south Mexico) and San Martin de Porres (east Bolivia) are in the lowlands characterized by distinct precipitation levels, resulting in neotropical forest and Chiquitano dry forest respectively (de la Vega-Leinert and Jiménez-Moreno in preparation). The Cantón Macará in south Ecuador is part of a semideciduous tropical dry forest ecosystem with unfavorable rainfall conditions due to high seasonality, annual variations, and the risk of extreme events (Kieslinger et al. 2019, Kieslinger Forthcoming). Finally, Nazareno in northeast Argentina is a transitional climatic and phytogeographic zone between Puna and Yungas with great environmental variety in different ecological floors (Steinhäuser 2020a, 2020b).

Nevertheless, our case studies illustrate distinct wider research contexts and backgrounds. In Tatahuicapan, the focus was on family farming production, household subsistence and food security strategies, household members migratory histories, young people's opinions on local opportunities, and the future of farming and food security. Palenque and San Martin de Porres were treated together comparing past, present, and future perceptions of the countryside and agriculture; local livelihoods; and youth perspectives in rural areas. In Macará, the main topics were the narratives about living in the countryside, in the city, and abroad; assessment of living conditions at the place of residence; negotiations of leaving and staying; and future visions for rural places. Finally, in Nazareno, the research background was landscape perception, connection to territory as Mother Earth, as well as strategies for building indigenous knowledge for future generations.

In line with the theoretical considerations of this paper, we did not use a fixed or bounded category of youth in our field work. The young people who participated in our research were approached differently according to the research context of the respective case studies. Further, we were interested in grassroot views and meaningful actions of the participants and established research as a joint process of knowledge production (Bergold and Thomas 2012). Thus, the methods chosen were applied flexibly (Beazley and Ennew 2014) and adapted to the participants in the research process. Consequently, our case studies used different formats for data collection and combined a variety of tools for data analysis as listed in the following (for detailed information, see the respective references).

- Tatahuicapan: Three field visits (2018–2019), sampling through local key informants and snowballing. Participatory workshops; open interviews with key actors; structured questionnaire applied in farmers' households; semistructured interviews conducted with farmers' sons and daughters. Two workshops N = 26 young people (16 to 29 years old); key actors N = 7 (local authorities, teachers, social organizations' representatives); farmer households N = 48; farmers' sons and daughters N = 45 (15 to 29 years old); coding after framework analysis; descriptive statistics; qualitative data analysis on historical land use changes at household scale (Jiménez-Moreno Forthcoming).

• Palenque and San Martin de Porres: Four field visits (2015–2018), sampling through local key informants and snowballing. Workshops using drawing and acting as catalyst for discussion; nonparticipant observation; field notes; collection of photos and drawings, voice recording. Palenque workshop: N = 45 pupils of secondary school (13 to 17 years old), and students of rural university (18 to 22 years old) of indigenous (first- generation city dweller) or mestizo ethnicities. San Martin de Porres workshop: N = 45 pupils of secondary school (13 to 17 years old) of Quechua ethnicity (first- or second-generation colonists); qualitative and visual content analysis; coding after Grounded Theory; computer supported analysis (R-Based Qualitative Data Analysis for Linux) (de la Vega-Leinert and Jiménez-Moreno in preparation).

• Macará: Field visit (2017) with two stage cluster sampling: Three workshops: N = 40 students (14 to 18 years old) of three classes of secondary schools in rural parroquias; survey: N = 282 students (14 to 20 years old), of selected graduation classes of secondary schools in the whole canton. Workshops: group discussions and visual tools; field notes, collection of photos, drawings and writings, voice recording; coding after Grounded Theory, qualitative content analysis, joint analysis of visual and textual data outputs. Surveys: structured standardized face-to face interviews, questionnaire with open questions; inductive coding and quantitative analysis with IBM SPSS Statistics 25; descriptive statistics (Kieslinger Forthcoming).

• Nazareno: Three field visits (2015–2017), sampling through local key informants and snowballing, for the present article the data of the young people was extracted. All interviewees identified themselves as indigenous. Walking interviews performed as "go along": 8 young adults (age range 30 to 40), peasants, students; ethnographic deep interviews and occasional encounters; participant observation; fieldnotes and research diary; collection of photos and drawings. Attendance of classes in secondary school: biology and earth sciences, 25 pupils each (14 to 17 years old); group discussion with students in the intercultural bilingual college: 8 students (25 to 30 years old). Coding after Grounded Theory; computer supported analysis with atlas.ti (Steinhäuser 2020a, 2020b).

In this paper, we analyzed the similarities, differences, and complementarities arising from our case studies to develop a joint reflection on the discussed topics. We triangulated visual and textual data outputs. Following a cyclical research process, we defined common questions, which guided the compilation of our data (see Charmaz 2014). This helped us to grasp the diversity of our participants' views, while keeping the contextuality of their respective social experiences in mind. Since we have worked on different phenomena, on different conceptual backgrounds, as well as specific local levels and settings, it would have been misleading to use a uniform category system. We chose to represent both commonalities and multifaceted richness in our empirical data by constructing a cluster of statements most relevant for participating youths (adopting their diverse wording) and our elaborated core codes. Only at the very end we translated the results and quotes from Spanish to English to avoid bias in the coding and analysis processes.

In the following sections, we first focus on young people's past and present projections on life in the countryside and in other places. Furthermore, we present motivations to stay or leave, migration decisions from the perspective of rural livelihoods, and associated impacts. We conclude with our participants' crosscutting visions for the countryside from all case studies.

Perceptions of Life in Rural Places in Relation to Other Places

We aimed to investigate how spatial and temporal dimensions affect perception of rural places, as young people's decisions to stay or leave are related to the way they perceive the place where they currently live. To this end, workshop participants in Palenque and San Martin de Porres were asked to draw and/or enact their perceptions of past life in the countryside, as told by their parents or grandparents. While in Palenque most participants were either born in town or had come to town to pursue their studies, all had parents or grandparents who still lived in the countryside (rural to urban mobility to enable education). Often, they were the first generation in their family to access secondary or university education. In San Martin de Porres, in contrast, all participants' (grand) parents originated from the Andes, and had emigrated to the lowland region (a rural-to-rural mobility often driven by extreme poverty and climate hazards in the emitting region and facilitated by land allocation during the implementation of the agrarian reform).

Past life was drawn, enacted, and described based on two main, partly complementary, partly contradictory narratives. Thus, participants in Palenque often started by emphasizing how past life had been in harmony with nature, centered on traditional, subsistence agriculture as the basis for healthy livelihoods and environments (Figure 2). Participants also identified several challenges, including the precariousness and hardship of life, poverty, lack of basic services (in particular health and education), and perspectives, as well as strict, repressive cultural conventions. Participants recurrently insisted that, while in former times farmers could live from subsistence agriculture, decades of neglect in public policies had led to its bankruptcy. In contrast, participants in San Martin de Porres first enacted the difficulties their parents and grandparents encountered to find land, adapt to lowland environmental conditions, establish their first settlements, organize their daily lives (how the men cleared the first agricultural plots, while the women cooked for the community), and stay in contact with their faraway families (Figure 3). They then emphasized how migration had enabled access to a land of hope, a new beginning for their families, and the opportunity to rise out of poverty and ascend socially.

FIG. 2—Picturing past life in the countryside (Palenque) Photograph taken in September 2017.

In Nazareno, where we aimed at understanding reasons for staying in such a remote location, young participant's answers reflect the values of the Andean cosmovision inherited by their ancestors. These values are constitutive for their sense of belonging, which is often constructed in contraposition to the city. The landscape is described in terms of esthetics, fertility, silence, and tranquility. The thankfulness for Earth's gifts was described in this way: "Here everything matures very nicely." This was strongly associated with healthy food for subsistence. In the words of one peasant woman, in the city "you have to work every day to eat, and you can't sow" (as agriculture is not considered work). However, the enjoyment of this—in their words "living with Mother Earth"—comes along with several challenges, namely the isolation, difficult access, hard work in the fields and with the animals, as well as dependency on weather conditions. Hardness of adaptation to life cycles of nature were particularly

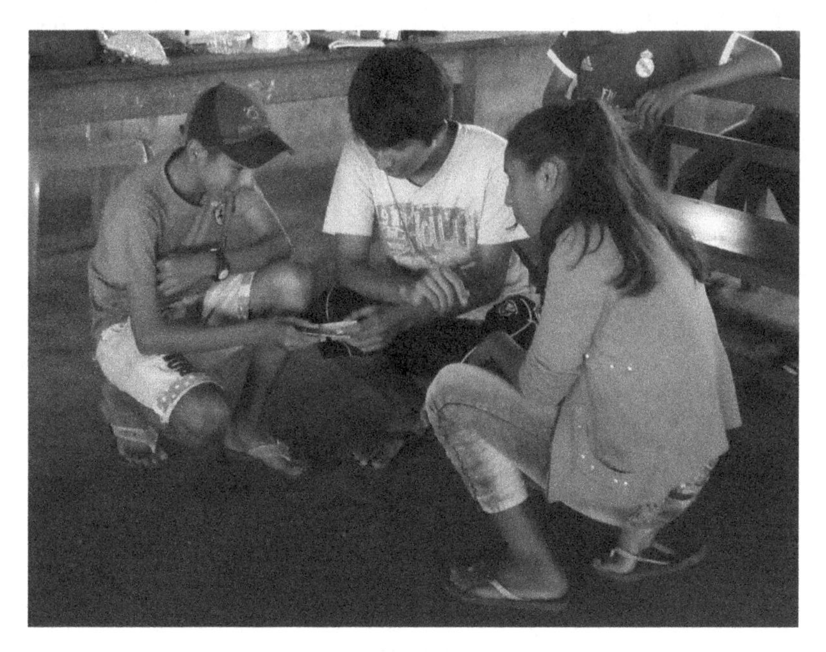

FIG. 3—Pupils enacting the arrival of a long-awaited letter from relatives who stayed in the Andes (San Martin de Porres) Photographs taken in April 2018.

pointed out by the young when compared to older participants of the community.

The greatest discrepancy in responses among young respondents in Nazareno was related to the advantages and disadvantages of isolation. One of them expresses it in this way: "You have nowhere to escape to here." Indeed, those who lived primarily in Nazareno pointed out more explicitly negative aspects of physical and social isolation. On the contrary, those who had lived long outside Nazareno described isolation as an opportunity to develop the values of Andean cosmovision (including reciprocity, living with Mother Earth, mindfulness), without interference from "modern" urban society, which is perceived as accelerated, consumerist, and superficial.

In Macará, we undertook a survey on young people's opinions about living conditions at their place of residence regarding personal development. Of a total of 282 survey participants, 278 provided multiple responses, with more abundant positive or neutral statements (n = 327) than negative ones (n = 244). As shown in Figure 4, statements about "life as a farmer/country life" and "nature/environmental services" were overall positive or neutral and issues related to the "conditions at place, community life, public services, and recreation/entertainment" included more positive or neutral affirmations than negative ones.

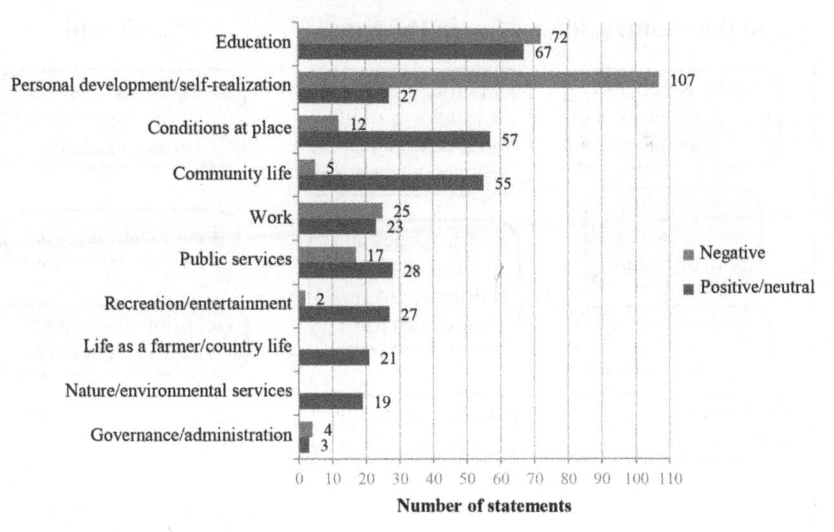

FIG. 4—Evaluation of living conditions at the place of residence in Macará.

Predominantly negative statements were given regarding "personal development/self-realization, education, work and governance/administration."

By means of participatory workshops, we grasped young people's narratives about life at different places. Figure 5 depicts their expressions taken from writings and subsequent group discussions. Interestingly, ways of living in the countryside were negotiated relationally toward cities and abroad. For instance, the participants associated the countryside with strong social ties and life with nature, but also deprivation, backwardness, and lack of opportunities for personal development. In contrast, the city offered opportunities to improve life, gain independence and self-realization, but also implied increasing living costs, contamination, and danger. Living abroad would offer better living conditions, in general, but also entail major hardships, like finding work, securing basic resources, or "becoming somebody." The affirmations indicated positive and negative aspects as well as connotations in between, which were also negotiated relationally. This becomes manifest in statements like "[the countryside] is relaxing, but you don't have all the comforts," "[in the city] there are opportunities, but it can be dangerous," and "[abroad] there are more working opportunities, but life is difficult there." These narratives revealed meanings of places, which may influence decision making over leaving and staying, but also on destinations of migration.

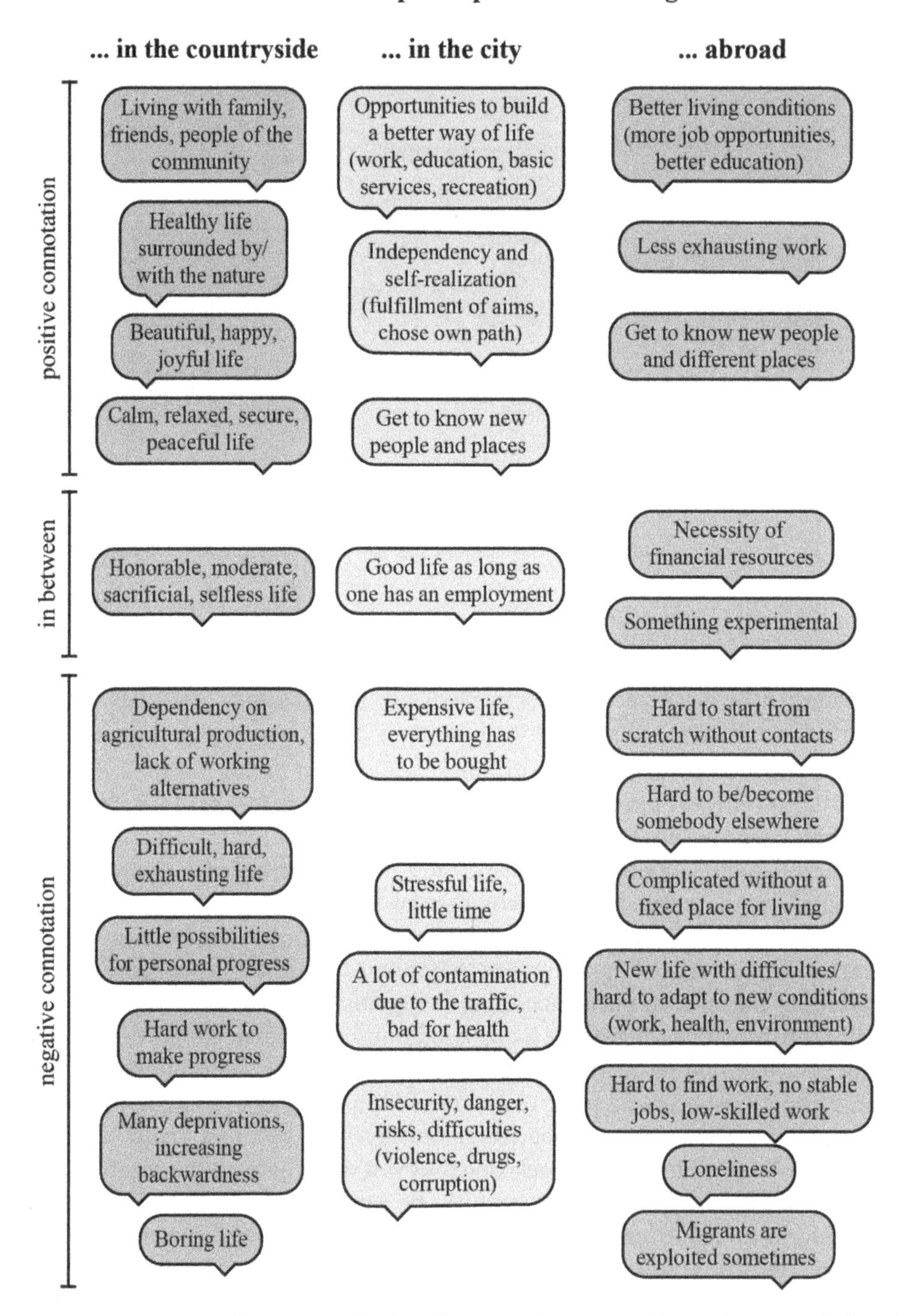

FIG. 5—Narratives of young people about living in the countryside, in the city and abroad (Macará).

Leaving and Staying at Rural Places

We were interested in young people's decision making on leaving and staying because their motivations fostered understanding of underlying meanings of places but also of desired identities and aspirations. In Macará, survey results show that young people are potentially highly mobile, as the majority of 275 respondents (98 percent—n = 280) aim to leave. The principal destinations indicated are cities in Ecuador (60 percent), such as Quito, Loja, or Guayaquil, and to a lesser-degree foreign countries (32 percent), principally the United States and Spain. The motivations for leaving (multiple responses n = 592) referred to education (31 percent), personal development and self-realization (25 percent), work (24 percent), conditions at places (12 percent), and social and family relations (7 percent). Importantly, the most frequent denomination of motivations varied according to the destinations (cities: education; foreign countries: work). The five students who aimed to stay (multiple responses n = 11) specified "being with the family" (5), "attractive place" (1), "priority of basic education" (1), "possibility of distance learning" (1), and "possibility of temporal movements" (1) as reasons.

In Nazareno, interviewees had had various mobility experiences, as temporary and seasonal migration to attend secondary school (before the opening of the Nazarene school), to accompany their parents for harvesting in other regions (during the summer school break), or for transhumance (to the valleys on the other side of the mountains). Reasons for remaining in Nazareno as identified in the deep interviews and participant observation were closely related to the search for a shared identity and motivated by a strong empathy for Mother Earth. A core element of their identity, subsistence farming, was described by young peasants during the walking interviews as an activity that required access to land, but also a high degree of practical experience (learning by observation), physical effort (tilling the land), perseverance (keeping the fields free of stones), and willpower. Interviewees were aware that many of these aspects of identity and community building had been forgotten or were no longer practiced. A young peasant concluded: "Today some people do not need to sow anymore to survive." Consequently, to keep young people in place a teacher explains how he encourages children to "affirm our identity firmly from a young age."

Migration and Rural Livelihoods

In Tatahuicapan, we explored the relationships between young people's migration and security strategies at household scale, which refer to ways families access food as part of their daily lives, including cultivation—either on own or leased land—and purchase with income acquired by diverse means. Additionally, we examined changes in economic and labor inputs from young migrants and possible influence on the use of family plots. Youth migration, categorized into educational and labor migration, was our first focus. Educational migration was usually practiced by young people belonging to families with a comparatively high, local socioeconomic status. Far from generating income and surplus that could be sent back in the form of remittances, this form of migration implies primarily a high economic cost for families, while young migrants generally do not come back to their home community to seek employment. In contrast, labor migration, primarily to regional and national destinations, is multifaceted. Most frequent types are seasonal agricultural laborers in major Mexican agro-industrial regions, construction workers and low-skilled services sector employees in regional and national urban destinations, low-skilled workers in manufacturing industries, and skilled professionals in different cities across the country. Labor migrants may or may not send remittances to their families according to their specific situations and sectors of employment.

Four food security strategies were identified in migrant-sending households (Figure 6), centered on the importance of food production with respect to food purchase, youth migration category, and young migrants' economic and labor inputs into the household economy, and family farming. These strategies showed a growing importance of food purchase—frequently dependent on government's welfare programs—over own production, as well as a rather scarce articulation of young people's mobility with family farming demands.

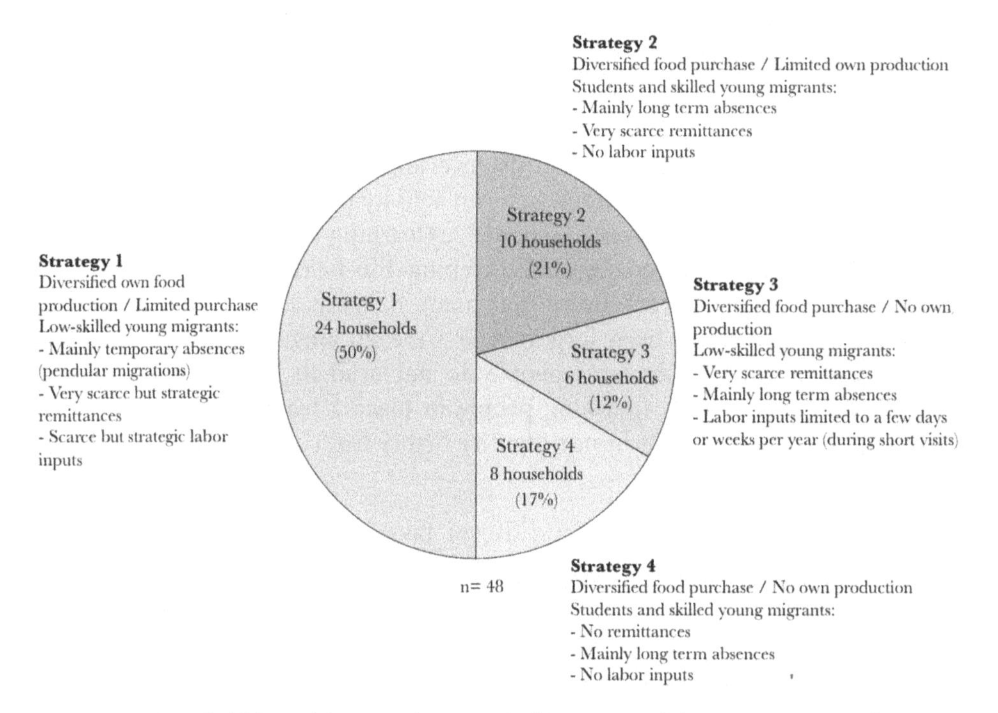

FIG. 6—Households' Food Security Strategies and Young People's Migration in Tatahuicapan.

For the most part, youth out-migration in Tatahuicapan does not imply a significant loss of labor force, because local farming practices require few workers, and farming laborers are in oversupply. Furthermore, young people's labor inputs tend to be scarce whether they migrate or not, which points to a generational lack of involvement with family farming that should be understood beyond out-migration per se. Indeed, in many households, attending school is perceived by young and old as the main responsibility of children and young people, who are thus less solicited to perform farm activities. This process, over the long term, affects youth's perception of, and engagement with farming, favoring their stepping out of such activities.

In most households investigated, youth migration did not bring systematic remittances, which were not considered an indispensable household income. Indeed, young people perceived savings generated through migration as a strategy to reach independence and realize individual projects, less so to invest in their parents' household and plots.

Young people who enrolled in professional studies rarely choose curricula related to farming or intended to improve family farming in the long term. Nevertheless, many interviewees perceived the knowledge and skills they acquired via educational or labor mobility as an advantage for them and their families, should they succeed in locally capitalizing these competences. However, interviewees, both young and old, identified several obstacles in the way, including the lack of access to land, credit, and technical and financial advice, and the shortage of well-paid local employment. These challenges restrict the economic and human resources young people may derive from migration, while they contribute to the continuous erosion of peasant identity.

Visions for the Future from Well-being to Bankruptcy

s a first step to explore how young people project themselves in specific places in the future, we examined how they perceived and anticipated change in rural places. Thus, when asked to portray the future of the countryside, pupils and students in Palenque pictured three broad trajectories of future change. First, some participants viewed a future in which the countryside would cater for the well-being of both people and nature, through the implementation of agroecological systems and sustainable technology. Second, some participants envisaged the abandonment and destruction of the current countryside, associated with the continuing bankruptcy of traditional peasant livelihoods and its replacement by an industrialized and urbanized countryside. Finally, other participants depicted a countryside, which successfully incorporated conventional intensive agriculture and cattle ranching and managed to create added-value products for sale, while it provided basic services and recreation (Figure 7). The latter is in keeping with the joint vision of participating pupils and students in San Martin de Porres, who

FIG. 7—Three visions of the future of the countryside (Palenque workshop): a) an environmentally-friendly countryside; b) a modernized countryside; c) a countryside turned city Photographs taken in September 2017.

depicted a countryside, which should offer them professional opportunities, economic growth through successful market integration, and access to modern services and sustainable technology. This is largely associated with the recent history of settlement and colonization in this dynamic frontier area, where agriculture is expanding and improving local living conditions, albeit at the cost of forested areas (de la Vega-Leinert 2020).

For young interviewees in Nazareno, the gradual landscape change they observed was closely related to changing lifestyles and eroding sense of community. Remembering her childhood and her games, one student stated: "Look, all the surroundings that you see. The landscape is beautiful and [people used to say]: 'it belongs to everyone,' and now you will see that it is fenced ... It hurts me to see that sharp line that cuts the hill in pieces." It was noticeable that topics related to the environment stood out most in the young people's narratives. While elders spoke mostly of their ancestral bond with Mother Earth, young people also pointed out the environmental problems generated by modern lifestyles, as indicated by the sites they chose for the walking interviews, such as a garbage dump and sites with visible overexploitation. This also became evident in drawings of participant secondary-school pupils (Figure 8) and the narratives that described them.

Different resources in the surroundings of Nazareno are commonly integrated into daily life, such as local building materials (mud, stones, straw, wood), water, or medicinal herbs. According to young people, this unrestricted use leads to depletion of resources. As a strategy for future education, the Intercultural Bilingual Education Center of Nazareno was created. Here, local students become teachers and recover the Andean cosmovision, the Quechua language, and regional handicrafts next to computer skills. Young people's narratives often departed from the traditional Andean indigenous cosmovision to merge with global awareness on sustainability.

FIG. 8—The "development of Nazareno in the next 15 years", showing waste disposal as one of the challenges (Participant observation in the "Earth Science" class, April 2016).

For example, during a group discussion, teacher students highlighted the responsibility of modern societies in causing climate change and identified some of its impacts regionally (melting of glaciers and increasing water scarcity). Thereby they reflected on their own responsibility and the uncertainties on how to adapt the irrigation practices, for example, as expressed by one student: "When sowing time begins, there is not enough water for all the fields."

Some or all these visions of the future were also found in the other two case studies, that we do not elaborate on for reasons of space. In Tatahuicapan, workshop participants envisaged a decaying vs. a modernized countryside. These views allocated very different meanings to agriculture: in the former vision it was associated with a life in poverty and, in the latter, with a profitable activity that guaranteed people's access to good quality, locally produced food. Finally, in Macará, young people imagined their living places either as devoid of perspectives, stating that "there is no future here," or as "a place with improvements." While some students portrayed increasing out-migration of people due to hard living conditions without amenities, others emphasized the responsibility of local people to improve living conditions at place, albeit with governmental support. For them, the countryside was associated with an urbanized and modernized place, but also with agricultural production, values of nature, and community.

Young People's Role in Future Developments

We discuss common insights emerging from our case studies in the form of a cluster as a basis for future comparative studies on youth perspectives on the countryside. Although portraying the plurality and multidimensionality of young

FIG. 9—Clustered view of the most relevant topics in the four case studies.

people's experiences and of their rural realities is beyond the limited scope of the present paper, Figure 9 depicts a diversity of coexisting topics arising from participating young people. These relate to current socio-ecological transformation processes affecting agriculture (such as integrate alternative networks of food), economic alternatives (like off-farm labor), population developments (for example, local communities incorporate external behaviors and values), and structural conditions (such as new technologies).

We placed participants' individual claims for more opportunities in the center of this cluster. These may be reinforced and contrasted by views in wider society (family, household, communities), an issue which is closely related to decision-making power and the ability to act and reflect underlying power imbalances (for example, related to gender). The narratives of our participants indicate individual evaluations, which are temporarily contextualized and fluid. Consequently, the topics may be (re)evaluated differently over the course of time.

Young people's multifaceted experiences are intertwined with a complex, often ambivalent, relationship toward the countryside, and personal future projections therein. The diversity of experiences illustrates how the common insights emerging in discussion with young people in our case studies cannot be treated as generalizations on hypothetical, homogeneous "rural young people," nor considered as linear cause-effect relations.

Our participants related spatial (im)mobilities and social and economic mobility, in the sense of "progressing" and "improving oneself" and "becoming somebody in life," which was also observed in the Latin American context by

Crivello (2015) and Diana Mata-Codesal (2015). The distinct motivations of young people to leave and stay overlap with their future visions of rural areas, which would offer better opportunities. One transversal topic, discussed by the young people in all case studies is the role of higher education in getting a better life—albeit with distinct meanings, expressions, and outcomes. Accessing higher education by out-migration can be perceived as a way out of agriculture and a means to socially ascend through off-farm labor. Also, it can open opportunities for young people in rural areas to generate new approaches to agriculture and new sources of employment and income in situ. From another perspective, facilities for higher education should, instead, be brought to the countryside (such as rural colleges and digital media) to provide young people access to these opportunities at place.

Our participants also pointed to intergenerational conflicts that may emerge due to competing individual aspirations and household or family goals. Accordingly, there are collective strategies that determine who leaves or stays, and the respective tasks for the former (acquiring education/employment, providing remittance) and the latter (caring for elderly, working the land). As depicted by Stockdale et al. (2018, 1), discourses on migration have often devalued staying in rural places and conceived immobility as staying behind or a failure to leave. In line with this perspective, several studies have shown how parents and communities legitimize and facilitate the need of mobility to make progress (Stockdale et al. 2018). Our case studies show ambivalence in this respect, where family members encourage young people to leave, but also to stay at place. Despite young people's experiences, a gap between their aspirations of (im)mobilities, choices, and the ability to act (see also desire-ability model of Carling 2002), in our case studies, staying does not mean immobility either (Barcus and Werner 2016; Morse and Mudgett 2017). Staying was an active attitude of some young people to develop the endogenous potential for the future of their place and live a better life than that proposed by the city. In this conscious decision of, in their words "rescuing the roots," they strengthen their links with the environment in which they live, and seek to recover and create the knowledge that allows them to do so.

Our case studies depict rural areas that are experiencing profound socio-ecological change against a backdrop of globalization and transforming agricultural communities through mobility, among other factors. Young people are actively participating in this multifaceted transformation process, thereby altering their perceptions and aspirations concerning the countryside. Further, they are exploring ways to shape their future, and their countryside, beyond, but often still strongly influenced by, the dichotomy of staying or leaving. As shown in our cluster, young people's visions do not evoke discrete, single pathways of rural livelihoods' transformation unlike, for example, the "hanging in, stepping up or stepping out" approach, proposed by Dorward et al. (2009). In contrast, they portray a wide and overlapping variety of possibilities that bring together the preservation and transformation of specific elements of their rural realities, and suggest an "openness" to be continually reassessed in response to material and symbolic changes within the countryside and abroad.

Based on our common results, we structured young people's future visions along a continuum. At one end, "living in harmony with nature" emphasizes the countryside as a place, where life is fulfilling and holistic, and where individuals are committed to supporting their community, preserving their cultural identity, and fostering sustainable life. At the other end, "living in the countryside as a city" symbolizes a progressively urbanized, industrialized, and globalized countryside, where young people can access basic services and integrate the mainstream society, albeit often in a marginal situation.

But these visions stress a dichotomic view that does not capture young people's projections of living elsewhere (Salazar 2011, for transnational context), an "elsewhere" that may not necessarily be away from the countryside, but rather another kind of countryside. We chose to name the final vision "living in between" to encapsulate a range of multifaceted, hybrid possible countrysides along this continuum. Here, young people reinvent the countryside by mingling different aspects, grouped in the other visions, according to their aspirations and preferences, including, for instance, access to innovative technology, but also the freedom to stay, move, and return. This can have different implications on the future role of agriculture in local livelihoods, and its form, labor opportunities for young people, land use, and relationships to Mother Earth. In this vision, young people may reclaim their cultural identity, strengthen their connection to their countryside, and (re)gain the power to develop their own transformation paths. A "better future in the countryside" as "living in between" may thus involve respecting Mother Earth, nurturing local food systems, but may also relate to a successful integration into commercial agriculture, which may or may not be fully compatible with, for example, sustainable agricultural transformations as expressed in current scientific and political debates on transition to agroecology (HLPE of the UN CFS [High Level Panel of Experts on Food Security and Nutrition of the Committee on World Food Security] 2019). A key challenge is how can young people contribute to shaping the countryside, which can fulfil their needs and aspirations, while taking responsibility for its sustainable use.

Conclusions

We brought together the perspectives of young people from four case studies in rural Latin America, which focused on understanding their visions on the countryside, and social and ecological transformations in the context of sustainability. Despite the different regional contexts and conceptual framings of our research, we identified many coincidences. Our participants' visions for life in the countryside can be described as hybrids that merge aspects of ways of living in rural and urban places, and are affected by imaginations of life in other places. The "living in between" vision takes multiple forms in a continuum and is fluid in place and time. Further, social meanings and experiences of staying and/or leaving depend on structural conditions, necessities, opportunities, and power to decide. From our participants' perspectives, the decision whether to leave or stay also depends on the access to resources, such as land; the ascribed meaning of local practices, for example traditional agriculture as hard work, and of identities: being a peasant can be associated culturally with belonging to a community but also with backwardness and poverty.

Both visions of rural places and decisions on leaving and staying affect local practices and the environment. Young people's mobilities are closely associated with a range of local socio-environmental implications, productive matrix, food security, and labor in the local economies of emitting communities, which may have significant impacts on those who are "left behind" (Torres and Carte 2016). Both rural development and migratory policies, therefore, have profound consequences for the future of young people as they significantly constrain them in their decision over which places they may live and work in (Bada and Fox 2021, for Mexico). We contend that these should be investigated together to explore ways to better coordinate joint approaches toward enabling young people's right to leave, but also to stay. This is, in our opinion, a cornerstone in ensuring fulfilling lives and the vitality of rural livelihoods.

Science and policy that aim at fostering sustainability pathways need to embrace young people's requirements, aspirations, and visions, and support them both in developing their countrysides and enabling their (im)mobilities through broader interconnected communities. Indeed, rural young people still have few possibilities to actively participate in shaping global discourses and future research on matters that determine their everyday life and constrain their futures. Based on the insights gained from our case studies, we recommend investigating how young people may be supported in their concerns about the countryside they want, as well as which arenas and processes may be most appropriate for young people to articulate their needs and demands. Future research should examine how the countryside may be transformed to maintain local identity, preserve landscape diversity, while opening perspectives for young people; a core issue here being the role young people's (im) mobilities may play in fostering hybrid countrysides toward sustainable development. Finally, we contend that science, policy, and development organizations must critically reflect on ways not to assign young people in rural areas preconceived roles that might limit their freedom of choice and agency.

Acknowledgments

Special thanks go to the young people and their communities who generously shared their experiences with us. Cristina de la Vega-Leinert is particularly grateful to Cristina García Ángel and Edmundo Gómez Horta and Isabel Mamani. Many thanks from Julia

Kieslinger to the local co-investigators Ángel Hualpa Erazo and Paola Rengel Vega who actively took part in the field work. Marcela Jiménez-Moreno is especially thankful to the people of Tatahuicapan, and to Ismael Arce Estrada and Esteban E. Ramírez Cruz for their valuable help during fieldwork. Cornelia Steinhäuser sincerely thanks Barbara Göbel, Gabriel Barba and Hugo Díaz Cabana for guidance and access to the community in its reciprocity with Mother Earth. Finally, we appreciate the comments of the two anonymous reviewers who contributed to more structure and clarity in our text.

Funding

A. Cristina de la Vega-Leinert was funded by the German Research Foundation under Project Grant Nr. VE 659/2-1. The research of Marcela Jiménez-Moreno was partially funded by UNAM-PAPIIT (project IN304519) and CONACyT (doctoral grant 298769). The research fieldwork of Cornelia Steinhäuser was funded by the international program IP@WWU of the German Academic Exchange Service (DAAD).

Notes

[1] https://www.americas2019.uni-bonn.de/.

[2] With the term "countryside" we refer to the emic concept of "el campo" as it is used by the participants of our research.

[3] The Food and Agriculture Organization of the United Nations states that, "Food security exists when all people, at all times, have physical and economic access to sufficient safe and nutritious food that meets their dietary needs and food preferences for an active and healthy life." World Food Summit 1996, Rome Declaration on World Food Security, http://www.fao.org/3/w3613e/w3613e00.htm. This definition characterizes food security with four dimensions: availability, access, use, and stability.

Orcid

A. Cristina De La Vega-Leinert http://orcid.org/0000-0003-0290-249X

Julia Kieslinger http://orcid.org/0000-0001-6584-6902

Marcela Jiménez-Moreno http://orcid.org/0000-0002-7723-2702

Cornelia Steinhäuser http://orcid.org/0000-0001-5744-8904

References

Aitken, S. C. 2001. *Geographies of Young People: The Morally Contested Spaces of Identity*. London: Routledge.

Appendini, K., and G. Torres-Mazuera. 2008. *¿Ruralidad sin Agricultura? Ciudad de*. México: El Colegio de México.

Bada, X., and J. Fox. 2021. Persistent Rurality in Mexico and 'The Right to Stay Home'. *The Journal of Peasant Studies*:1–25. doi:10.1080/03066150.2020.1864330.

Barcus, H., and C. Werner. 2016. Choosing to Stay: (Im)mobility Decisions Amongst Mongolia's Ethnic Kazakhs. *Globalizations* 14 (1):32–50. doi:10.1080/14747731.2016.1161038.

Bauman, Z. 1998. *Globalization: The Human Consequences*. Cambridge, U.K.: Polity.

Beazley, H., and J. Ennew. 2014. Participatory Methods and Approaches: Tackling the Two Tyrannies. In *Doing Development Research*, edited by V. Desai and R. B. Potter, 189–199. London: Sage.

Bell, M. M., and G. Osti. 2010. Mobilities and Ruralities: An Introduction. *Sociologia Ruralis* 50 (3):199–204. doi:10.1111/j.1467-9523.2010.00518.x.

Bergold, J., and S. Thomas. 2012. Participatory Research Methods: A Methodological Approach in Motion. *Forum: Qualitative Social Research* 13 (1):Art. 30.

Bushin, N. 2009. Researching Family Migration Decision-Making: A Children-in-Families Approach. *Population, Space Place* 15 (5):429–443. doi:10.1002/psp.522.
Carling, J. 2002. Migration in the Age of Involuntary Immobility: Theoretical Reflections and Cape Verdean Experiences. *Journal of Ethnic and Migration Studies* 28 (1):5–42. doi:10.1080/13691830120103912.
Carling, J., and K. Schewel. 2017. Revisiting Aspiration and Ability in International Migration. *Journal of Ethnic and Migration Studies* 44 (6):945–963. doi:10.1080/1369183X.2017.1384146.
Cazzuffi, C., and J. Fernández. 2018. *Rural Youth and Migration in Ecuador, Mexico and Peru*. Serie documento de trabajo N° 235. Programa Jóvenes Rurales, Territorios y Oportunidades: Una estrategia de diálogos de políticas. Santiago, Chile: Rimisp.
Chakraborty, K., and S. Thambiah. 2018. Children and Young People's Emotions of Migration across Asia. *Children's Geographies* 16 (6):583–590. doi:10.1080/14733285.2018.1503231.
Charmaz, K. 2014. *Constructing Grounded Theory: A Practical Guide Through Qualitative Analysis*. 2nd ed. Los Angeles: Sage.
Coulter, R., M. Van Ham, and A. M. Findlay. 2016. Re-Thinking Residential Mobility: Linking Lives through Time and Space. *Progress in Human Geography* 40 (3):352–374. doi:10.1177/0309132515575417.
Crivello, G. 2015. 'There's No Future Here': The Time and Place of Children's Migration Aspirations in Peru. *Geoforum* 62:38–46. doi:10.1016/j.geoforum.2015.03.016.
Dalsgaard Pedersen, H. 2018. Is Out of Sight Out of Mind? Place Attachment among Rural Youth Out-Migrants. *Sociologia Ruralis* 58 (3):684–704. doi:10.1111/soru.12214.
de la Vega-Leinert, A. C. 2020. Too Small to Count? Making Land Use Transformations in Chiquitano Communities of San Ignacio De Velasco, East Bolivia, Visible. *Journal of Land Use Science* 15 (2–3): 172–202. doi: 10.1080/1747423X.2020.1753834.
de la Vega-Leinert, A. C., and P. Clausing. 2016. Extractive Conservation: Peasant Agroecological Systems as New Frontiers of Exploitation? *Advances in Research* 7 (1):50–70.
de la Vega-Leinert, A. C. and M. Jiménez-Moreno. in preparation. ¿Cuál campo a futuro? Ambivalencia de Juventudes hacia el Campo. In Naturalezas Urbanizadas, Urbes Naturalizadas. Relaciones Socionaturales y Territorios Híbridos, edited by F. Figueroa and M.G. Guzmán Chávez.
Dorward, A., S. Anderson, Y. N. Bernal, E. S. Vera, J. Rushton, J. Pattison, and R. Paz. 2009. Hanging In, Stepping up and Stepping Out: Livelihood Aspirations and Strategies of the Poor. *Development in Practice* 19 (2):240–247. doi:10.1080/09614520802689535.
Fabregat, E., S. Vinyals-Mirabent, and M. Meyersm. 2020. "They are Our Brothers": The Migrant Caravan in the Diasporic Press. *Howard Journal of Communications* 31 (2):204–217. doi:10.1080/10646175.2019.1697400.
Farmer, D. 2017. Children and Youth's Mobile Journeys: Making Sense and Connections within Global Contexts. In *Movement, Mobilities, and Journeys, Geographies of Children and Young People* Vol. 6, edited by C. Ni Laoire, A. White, and T. Skelton, 245–269. Singapore: Springer.
Feixa, C. 1999. *De Jóvenes, Bandas y Tribus. Antropología de la Juventud*. Barcelona: Ariel.
Flick, U. 2018. *The SAGE Handbook of Qualitative Data Collection*. London: Sage.
Geiger, M., and M. Steinbrink. 2012. *Migration und Entwicklung*: Merging Fields in Geography. *IMIS-Beiträge* (42):7–36. https://repositorium.ub.uni-osnabrueck.de/bitstream/urn:nbn:de:gbv:700-2013091811577/1/imis42.pdf
Glick Schiller, N. 1997. From Immigrant to Transmigrant: Theorizing Transnational Migration. In *Transnationale Migration*, edited by L. Pries, 121–140. Baden-Baden: Nomos.
Guiskin, M., P. Yanes, and M. Del Castillo Negrete. 2019. *The Rural Youth Situation in Latin America and the Caribbean*. Rome: International Fund for Agricultural Development.
Hecht, S., A. L. Yang, B. Sijapati-Basnett, C. Padoch, and N. L. Peluso. 2015. *People in Motion*, Forests *in Transition: Trends in Migration*, Urbanization, *and* Remittances and Their Effects on Tropical Forests. CIFOR Occasional Paper 142. Bogor, Indonesia: Center for International Forestry Research.
HLPE of the UN CFS [High Level Panel of Experts on Food Security and Nutrition of the Committee on World Food Security]. 2019. *Agroecological and other Innovative Approaches for Sustainable Agriculture and Food Systems that Enhance Food Security and Nutrition*. Rome:

A Report by the High-Level Panel of Experts on Food Security and Nutrition of the Committee on World Food Security.

Huijsmans, R. 2017. Children and Young People in Migration: A Relational Approach. In *Movement, Mobilities, and Journeys*, edited by C. Ní Laoire, A. White, and T. Skelton, 45–66. Singapore: Springer.

Jiménez-Moreno, M. Forthcoming. Youth Outmigration in the Sierra De Santa Marta Region in Mexico: Implications for Local Food Security and Prospects for a Forest Transition. PhD diss. National Autonomus University of Mexico.

Kay, C. 2008. Reflections on Latin American Rural Studies in the Neoliberal Globalization Period: A New Rurality? *Development and Change* 39 (6):915–943. doi:10.1111/j.1467-7660.2008.00518.x.

Kieslinger, J. Forthcoming. The Importance of Spatial (Im)mobilities in the Context of Changing Life Conditions and Lifeworlds: The Example of Socio-Ecological Transformation in Rural Ecuador. PhD diss. Friedrich-Alexander Universität Erlangen-Nürnberg.

Kieslinger, J., P. Pohle, V. Buitrón, and T. Peters. 2019. Encounters between Experiences and Measurements: The Role of Local Knowledge in Climate Change Research. *Mountain Research and Development* 39 (2). doi:10.1659/MRD-JOURNAL-D-18-00063.1.

Kieslinger, J., S. Kordel, and T. Weidinger. 2020. Capturing Meanings of Place, Time and Social Interaction When Analyzing Human (Im)mobilities: Strengths and Challenges of the Application of (Im)mobility Biography. *Forum Qualitative Sozialforschung/Forum: Qualitative Social Research* 21 (2): Art. 7. doi:10.17169/fqs-21.2.3347.

Krauskopf, D. 2005. Desafíos en la Construcción e Implementación de las Políticas de Juventud en América Latina. *Nueva Sociedad* 200:141–153.

Low, S. M., and I. Altman. 1992. Place Attachment: A Conceptual Inquiry. In *Place Attachment*, edited by I. Altman and S. M. Low, 1–12. New York: Plenum.

Massey, D. S., J. Arango, G. Hugo, A. Kouaouci, A. Pellegrino, and J. E. Taylor. 1993. Theories of International Migration: A Review and Appraisal. *Population and Development Review* 19 (3):431. doi:10.2307/2938462.

Mata-Codesal, D. 2015. Ways of Staying Put in Ecuador: Social and Embodied Experiences of Mobility–Immobility Interactions. *Journal of Ethnic and Migration Studies* 41 (14):2274–2290. doi:10.1080/1369183X.2015.1053850.

Meyer, F. 2017. Navigating Aspirations and Expectations: Adolescents' Considerations of Outmigration from Rural Eastern Germany. *Journal of Ethnic and Migration Studies* 44 (6):1032–1049. doi:10.1080/1369183X.2017.1384163.

Milbourne, P., and L. Kitchen. 2014. Rural Mobilities: Connecting Movement and Fixity in Rural Places. *Journal of Rural Studies* 34:326–336. doi:10.1016/j.jrurstud.2014.01.004.

Morse, C. E., and J. Mudgett. 2017. Happy to Be Home: Place-Based Attachments, Family Ties, and Mobility among Rural Stayers. *The Professional Geographer* 70 (2):261–269. doi:10.1080/00330124.2017.1365309.

Murdoch, J. 2005. *Post-Structuralist Geography: A Guide to Relational Space*. London: Sage.

Ní Laoire, C., and A. Stockdale. 2016. Migration and the Life Course in Rural Settings. In *Routledge International Handbook of Rural Studies*, edited by M. Shucksmith, 36–49. New York, NY: Routledge.

Ní Laoire, C., F. Carpena-Méndez, N. Tyrrell, and A. White. 2010. Introduction: Childhood and Migration — Mobilities, Homes and Belongings. *Childhood* 17 (2):155–162. doi:10.1177/0907568210365463.

OECD [The Organisation for Economic Co-Operation and Development]. 2018. *The Future of Rural Youth in Developing Countries: Tapping the Potential of Local Value Chains. Development Centre Studies*. Paris: OECD Publishing.

Ostrom, E. 2000. Collective Action and the Evolution of Social Norms. *Journal of Economic Perspectives* 14 (3):137–158. doi:10.1257/jep.14.3.137.

Punch, S. 2001. 'Negotiating Autonomy: Childhoods in Rural Bolivia,' in Conceptualising Child-Adult Relations. In *The Future of Childhood Series*, edited by L. Alanen and B. Mayall, 23–36. London: Routledge.

Rae-Espinoza, H. 2016. Transnational Ties: Children's Reactions to Parental Emigration in Guayaquil, Ecuador. *Ethos* 44 (1):32–49. doi:10.1111/etho.12111.

Rodríguez, E. 2011. *Políticas de Juventud y Desarrollo Social en América Latina: Bases para la Construcción de Respuestas Integradas*. Foro de Ministros de Desarrollo Social de América, San Salvador. 11 July.

Salazar, N. B. 2011. The Power of Imagination in Transnational Mobilities. *Identities* 18 (6):576–598. doi:10.1080/1070289X.2011.672859.

Samers, M. 2010. *Migration*. New York: Routledge.

Schmuck, M. E. 2019. Juventudes en Plural, Territorios en Transformación. Hacia un Estado del Arte de los Estudios sobre Juventudes Rurales en Argentina. *PÓS* 14 (1):38–56.

Sheller, M., and J. Urry. 2006. The New Mobilities Paradigm. *Environment & Planning A* 38 (2):207–226. doi:10.1068/a37268.

Soja, E. 1989. *Postmodern Geographies. The Re-Assertion of Space in Critical Social Theory*. London: Verso.

Steinhäuser, C. 2020a. Mountain Farmers' Intangible Values Foster Agroecological Landscapes. Case Studies from Sierra Santa Victoria in Northwest Argentina and the Ladin Dolomites, Northern Italy. *Agroecology and Sustainable Food Systems* 44 (3):352–377. doi:10.1080/21683565.2019.1624285.

———. 2020b. Los Saberes de los Ancestros. Clave para los Vínculos con la Madre Tierra en una Comunidad Andina en Argentina. *Documents d'Anàlisi Geogràfica* 66 (2):307–324. doi:10.5565/rev/dag.606.

Stockdale, A., N. Theunissen, and T. Haartsen. 2018. Staying in A State of Flux: A Life Course Perspective on the Diverse Staying Processes of Rural Young Adults. *Population, Space Place* 24:e2139. doi:10.1002/psp.2139.

Torres, R. M., and L. Carte. 2016. Migration and Development? The Gendered Costs of Migration on Mexico's Rural "Left Behind". *Geographical Review* 106 (3):399–420. doi:10.1111/j.1931-0846.2016.12182.x.

UN [United Nations]. 2015. *General Assembly Resolution A/RES/70/1. Transforming Our World, the 2030 Agenda for Sustainable Development*. https://www.un.org/ga/search/view_doc.asp?symbol=A/RES/70/1&Lang=E.

———. 2018. *Youth and the 2030 Agenda for Sustainable Development*. New York: United Nations.

Urteaga, M., and L. F. García. 2015. Juventudes Étnicas Contemporáneas En Latinoamérica. *Cuicuilco* 22 (62):7–35.

Yeates, N. 2012. Global Care Chains: A State-of-the-Art Review and Future Directions in Care Transnationalization Research. *Global Networks* 12 (2):135–154. doi:10.1111/j.1471-0374.2012.00344.x.

Yuval-Davis, N. 2006. Belonging and the Politics of Belonging. *Patterns of Prejudice* 40 (3):197–214. doi:10.1080/00313220600769331.

SECOND HOME PROPERTY OWNERSHIP AND PUBLIC-SCHOOL FUNDING IN WISCONSIN'S NORTHWOODS

RYAN DOUGLAS WEICHELT and EZRA ZEITLER

ABSTRACT. Northern Wisconsin's tourism economy has drawn the attention of scholars interested in the economic linkages that exist between it and sending areas in the urban and suburban Upper Midwest. Interrelationships between the two areas have left an indelible mark on the rural landscape through the presence of lakefront vacation homes owned by nonresidents. Where are these properties located, and where do the nonresident landowners call home? Geospatial data for 816,000 parcels, provided by the Wisconsin State Cartographer's Office, are utilized to reveal the extent and value of nonresident property in the Northwoods. Schumpeter's theory of Creative Destruction is used as a framework to explain the emergence of tourism-dependent communities in the region, and case studies reveal the impacts of nonresident property taxes on public school funding. These findings suggest that as population in the Northwoods declines in upcoming years, communities will increasingly rely on nonresident investment to sustain local institutions.

Although relatively unknown to many outside of the Midwest, Wisconsin's Northwoods is a popular tourism destination for the region's residents. This rural region offers access to thousands of acres of forested public land and more than 10,000 lakes for year-round activities. During the summer months, residents and tourists enjoy more than 1,152 miles of bike trails and hundreds of miles of all-terrain vehicle (ATV) and utility task vehicle (UTV) trails. Autumn lures weekend visitors seeking the bright colors of hardwood leaves before winter sets in. According to the Wisconsin Department of Natural Resources (2020), in winter months, thousands of miles of groomed trails attract snowmobile owners from throughout the region. Cross-country skiers enjoy hundreds of miles of dedicated trails, including those that host the largest race in the country, the American Birkebeiner, in February. These amenities are the centerpiece of a regional economy that is sustained by tourism. According to the Wisconsin Department of Tourism (2020), tourism activities in northern Wisconsin generate nearly 2 USD billion in spending.

The lakes and forests of northern Wisconsin have attracted pleasure-seekers from the Upper Midwest and beyond for generations. After treaties between the Federal government and the Ojibwe and Menominee in 1836, 1837, and 1842 opened the region to colonization and industrial development (Satz 1991; Loew 2013), economic activities entailed large-scale timber harvesting (Rohe 1997), localized mineral extraction (Alanen 1997; Liesch 2006), paper production (Weichelt 2016), and limited success in agriculture (Gough 1997). As old-growth forests were cleared and returns on investment declined at the turn of the twentieth century, county, state, and federal governments converted harvested forest land and abandoned agricultural lands into public forest lands that comprise between one-third to two-thirds of land in a majority of counties in this region today.

Louis Turner and John Ash (1975) coined the term "pleasure periphery" in relation to tourism. The authors defined these peripheral areas as playgrounds for urban dwellers to avoid the restraining responsibilities of their homes. As the forests of northern Wisconsin regrew in the early twentieth century, railroad companies collaborated with tourist organizations and local governments to market the natural beauty and recreational amenities of the region to wealthy urban residents of southern Wisconsin, northern Illinois, and the Twin Cities region of Minnesota (Shapiro 2013). The culmination of rising incomes, increasing leisure time, and better-quality roads after World War II improved access to the region for middle-class residents of southern Wisconsin, Minnesota, and northern Illinois (Bawden 1997; Hart 1984; Kates 2001).

Financial investments in land and second homes by many of these seasonal visitors helped establish multigenerational ties to northern Wisconsin that continue today. The high degree of place attachment to this loosely defined region, known colloquially as the "Northwoods," is reflected in the work of scholars (Lindmark 1971; Stedman 2003; 2006) and journalists based in southern Wisconsin who describe the tranquil natural environment of the Northwoods to their subscribers in the metropolitan regions of Madison (Adams 2014), Milwaukee (Lewis 2017; Jones 2018), and Chicago (Solomon 2004; Moran 2018).

The transformation of the Northwoods from a region where lumbering and mining were dominant economic activities into a tourism dependent region or "pleasure periphery" can be partially explained by Joseph Schumpeter's (1942) theory of Creative Destruction. Typically used to contextualize the transformation of primary- and secondary-sector landscapes into consumerist landscapes, creative destruction is a process in which capitalism's search for profit leads to evolutions in economic activities. Such is the case for many areas of the Northwoods, where economic activities associated with the extraction of natural resources evolved into the commoditization of engagement with natural amenities. Clare Mitchell (1998) argues that rural, amenity-rich places appeal to nostalgia-driven, postmodern urban and suburban consumers looking to invest in property where they can spend leisure time. The phenomenon is also partially contextualized by David Harvey, who argues that capitalism ultimately creates "rational landscapes" in which places function as the centers of capital accumulation (1985, 1988). Capital accumulation over the past century in the Northwoods can be observed here through nonresident acquisition of mainly lakefront property. As demand for lakefront property increased in the region, land values increased in kind. As supply declined, residents, who on average earned less than the average nonresident landowner, found themselves less able to purchase desirable lakefront property. Success in this regard can create social animosity between nonresident property owners and permanent residents over a fear that their idyllic rural community is being overrun by outsiders, and that local governments and businesses cater to their needs at the expense of local residents (Vanderwerf 2008). Mitchell (1998) argues that this situation may eventually lead to locals accepting that they no longer "own" their community, but are rather controlled by the consumer driven demands of nonlocals. Her study of St. John's, Newfoundland, identified the loss of control locals had on the community, leaving it, as she claims, in a state of disequilibrium due to increased influences of tourists. This process exists in Wisconsin's Northwoods, though it is not spatially uniform.

Despite the long economic dependence on tourism in the Northwoods, geographic studies examining where nonresident landowners permanently reside or where they own recreational property in the region are limited. John Fraser Hart (1984) identified seven significant resort areas at county and township levels in Wisconsin through the analysis of seasonally vacant housing data from the U. S. Census and gross sales receipts in the lodging sector provided by the State Department of Revenue. Four of the seven resort areas are in the northern third of the state. Dave Marcouiller and others (1996) conducted a similar study in the following decade and, using U.S. Census housing data to locate densities of recreational housing units at the county-level between 1970 and 1990, highlight a significant increase in the number of second homes in the Northwoods during this period. Due to technological limitations and the nature of how data were collected and agglomerated, detailed spatial analysis was unavailable in these studies. Advances in geographic information systems (GIS) and organizational management of geospatial land-management records, however, now make it possible to analyze patterns of second home ownership at the most local scale —individual parcels. In July 2015, the Wisconsin State Cartographer's Office (WSCO) released its first statewide GIS parcel-fabric database to the public. It contained more than 3.4 million land parcels and attribute data that includes parcel geometry, coordinates, size, the primary address of the landowner, and assessed fair market value (See 2017). In Wisconsin, tax bills are sent to residents in early December with the first payments due after. As landowner addresses in the parcel databases are used by county treasurers to send annual property tax bills in December, when many nonresident property owners are away, they are considered an indicator of primary or secondary property status in this study. This study is the first to utilize a statewide land-ownership database to analyze extensive spatial patterns of second home ownership at the parcel level in the United States.

Joshua Hagen quotes David Lowenthal: "The past as we know it is partially a product of the present; we continually reshape memory, rewrite history, and refashion relics" (2006, 3). In a globalized context, Dieter Müller (2005) argues globalization has created a world of high mobility, transforming the tourist landscape, in this case the Northwoods, into a commonly visited place with differing interpersonal interactions that are nonetheless familiar. Interactions with the Northwoods landscape through actual experiences, stories passed from family to family, or digitally through images via social media, justify the existence of the region. This study contributes to research in the realm of sustainability in rural places by addressing empirical questions and structural economic issues related to tourism in northern Wisconsin. The first approach of this analysis utilizes the WSCO parcel-fabric database to locate nonresident properties in the Northwoods and identify places where nonresident Northwoods property owners call home. With a study area that includes more than 816,000 individual parcels, what are the patterns of parcel ownership in the Northwoods? Are there patterns of ownership that are not only in Wisconsin, but extend regionally and nationally? Furthermore, do patterns of ownership provide some areas greater economic advantages compared to others?

To ascertain the economic impacts of second homeownership, a more specific analysis of four individual school districts will be utilized through the lens of property values and property taxes. Property taxes vary in how they are levied upon property owners throughout the United States due to differences in the types of services rendered by local governments. In Wisconsin, a significant proportion of public-school funding is derived from annual property taxes paid by the owners of parcels lying within school district boundaries. Specifically, this analysis will identify the extent to which local public-school funding benefitted from nonresident property owners during a period of declining state support. The districts chosen for this analysis vary in their appeal to outside tourists searching for large bodies of water to recreate, to winter landscapes providing snow for skiing or snowmobiling. Based on these desires to consume the natural landscape, do varying degrees of second homeownership represent a "rational landscape" in which prices are dictated by capital accumulation in amenity-rich locations? Furthermore, do unequal patterns of tourist amenities, as Mitchell identified in St. John's, possibly leave areas of the Northwoods in a state of disequilibrium through the forces of land holdings by nonlocals?

Methods

Determining the extent of the study area required conversations about the boundaries of the Northwoods region, as it is considered as much a vernacular region as a formal region. An academically intuitive definition would include the calculation of forest cover and population density, but it is common for people to use a highway intersection, river crossing, a vista of rolling hills covered in pine trees, or a roadside

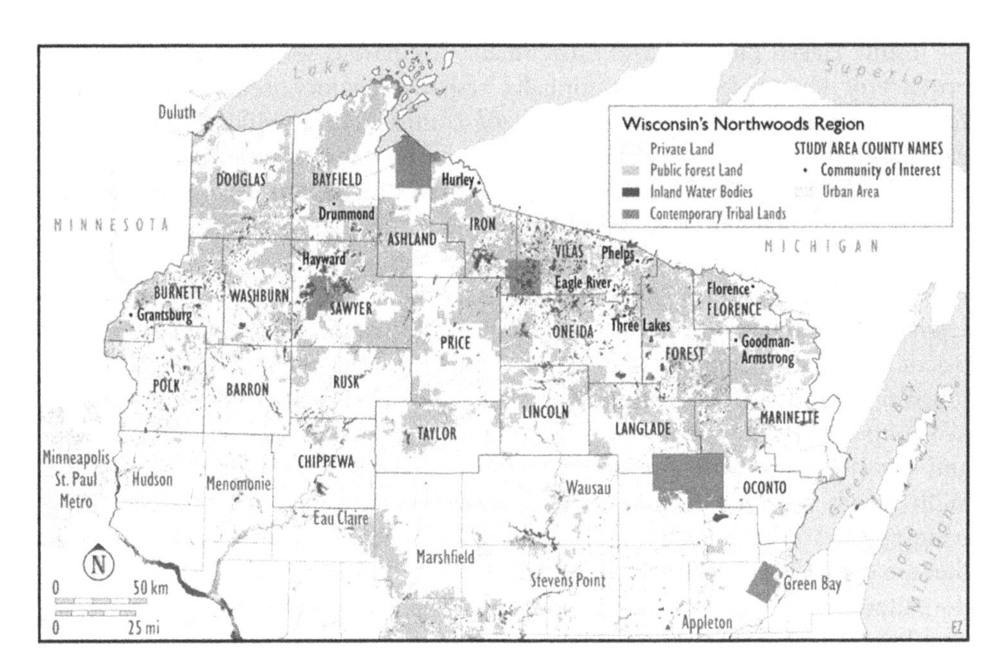

Fig. 1—Map of the Northwoods Region of northern Wisconsin.

bait shop as cognitive indicators that they have arrived (Nelson 2017). For the purposes of this study, the Northwoods includes twenty-one counties in the northern third of the state (Figure 1). Most of this vast, 21,000 square mile (13.44 million acre) region is covered with stands of mixed hardwood and softwood trees, concentrations of freshwater lakes attributed to the geologically recent Wisconsin glacial period, and pockets of agricultural land that remain from the cutover period of the nineteenth century. The western, southern, and eastern margins of the study area transition from heavily forested landcover commonplace in the interior to the agricultural landscape for which Wisconsin is recognized. The region's largest population centers, which vary in size from 10,000 to 25,000 residents, are in its margins. Sawyer and Washburn counties in the west and Oneida and Vilas counties in the north central portion of the region are home to a majority of the area's most-popular natural amenities.

The Wisconsin State Biennial Budget of 2013–2015 included Act 20, legislation mandating local governments to coordinate with the WSCO to develop a statewide digital parcel database. In a 2013 memo from the Wisconsin Department of Administration (DOA) Section 186 of Act 20, the goal was to develop a " … GIS that meets end users' business needs" (2013, 3). Sections 1247d, 1247h, and 1247p identified the data all local governments were instructed to include in the database. These data included the mailing address of the parcel owner, the assessed value of the land, the assessed value of improvements, the total assessed value, the class of property, the estimated fair-market value, the total property tax, and any acreage data pertaining to the parcel. Supplemental information in this data set included the local address of the parcel (if applicable), the school district in which the parcel is found, value of forest land (if applicable), the classification of the property, the auxiliary class of the property, and the latitude and longitude of the centroid of the parcel.

This study utilized Version 3 (V3 Project) data from the WSCO and was created from 2016 and 2017 data provided by local municipalities and counties as required by Act 20. The resulting database included 3.486 million parcels across the state (See 2017). Of interest to this study was the parcel tax mailing address. This address became the determining factor for classification of parcel owners. Yet due to the nature of how the address was identified (that is, 10 4th Ave., Shell Lake, WI 54871) in the database, additional transformations were necessary for further categorization and geocoding of parcel-owner addresses. In order to efficiently separate the elongated addresses into their respective fields, a multitude of Excel functions were used (i.e. = LEFT (A1, (FIND ("*", SUBSTITUTE (A1, " ", "*", LEN(A1)—LEN(SUBSTITUTE(A1, " ", ""))))))-4)], in conjunction with the column-wizard tool in Excel to correct the different address formats. These resulting transformations separated the address from the city, state, and zip code into individual columns, allowing for further categorization and geocoding. Individual addresses were geocoded using ESRI's Business Analyst Extension. This process yielded a success rate of greater than 99 percent. Addresses that could not be geocoded were usually due to an incorrectly transcribed zip code or misspelled street name. These errors were easily corrected.

The first geographic approach used in this study uses geocoded owner addresses to identify the extent of nonresident land ownership in the study area. ESRI's ArcMap GIS software provided the geoprocessing analysis tools (clip, spatial join) necessary to manage and analyze the large dataset at the county level. From this scale, parcel ownership, parcel size, and assessed value were extracted to calculate summary statistics to provide an overview of private land ownership in the region.

The second geographic approach focuses on the patterns of parcel ownership at a local district level. Due to the availability of parcel land value and collected taxes from individual parcels, it is possible to calculate the economic advantages this provides to local communities, especially school districts. The addresses in the data were utilized to categorize parcel owners. These categories include local owners living in Wisconsin (those living within the Northwoods region defined above), nonlocal parcel owners living in Wisconsin, owners from Minnesota, owners from Illinois, and other parcel owners both domestic and international. This categorization allowed for further economic analysis of how much land was owned by groups, the value of that land, and the subsequent taxes paid by each group. Furthermore, the categorization of landowners will also be mapped at a local scale to show patterns of ownership along highly valued lakefront properties.

Four Northwoods school districts—Grantsburg, Hayward, Hurley, and Northland Pines in Eagle River—were chosen for this analysis (Figure 1). These school districts represent communities across northern Wisconsin with varying concentrations of tourism amenities. For example, Hayward and Northland Pines school districts include the cities of Hayward and Eagle River, both areas with well-established, tourism-dependent economies. Both cities boast large lakes for fishing, boating, and other water-based activities, and provide seasonal opportunities for biking, hiking, skiing, and snowmobiling. A local analysis of the lakes near Hayward and Eagle River will provide a spatial representation of ownership along bodies of water. Contrasting this are the Grantsburg and Hurley school districts. These communities lack the abundance of lakes for which Hayward and Eagle River are known. A large portion of the Grantsburg School District is a state natural area (Reed Lake State Natural Area), which curtails human development. Though the Gile Flowage offers tourists in the Hurley School District opportunities to fish and boat, local land management forbids development along its shores. The city of Hurley does offer a gateway to the ski resorts of Northern Michigan and is located along byways for ATVs and snowmobiles, but overall it lacks the high-priced amenities many second homeowners are looking for as seen in places like Hayward and Eagle River.

Who Owns the Northwoods?

Geocoded WSCO parcel data reveal that approximately 36 percent of property in the study area is administered by local, state, and federal governments, and the remaining 64 percent consists of 718,808 privately owned parcels. Thousands of these parcels are owned by Georgia-Pacific, Enbridge, and other corporations, but a majority are owned by individuals or families and used in a permanent or seasonal residential capacity. Distinguishing these two residential categories is possible by considering the owner mailing address included in the data set. Because the address is used to send the property tax bill to the owner every December, it is reasonable to assume that the property is more likely to be a permanent residence if the address is local and nonlocal addresses are the primary residences of nonresident property owners.

Fifty-eight percent (419,583 of 718,808) of privately owned parcels in the study area can be considered local, as their address matches the owner mailing address. Their average size was 17.1 acres and average assessed value was 93,428 USD. Permanent residents of the study area who own additional properties in the study area own a total of 39,678, or 5.5 percent, of the parcels. With an average size of 25 acres and assessed value of 72,679 USD, these parcels tend to be undeveloped and located away from water bodies.

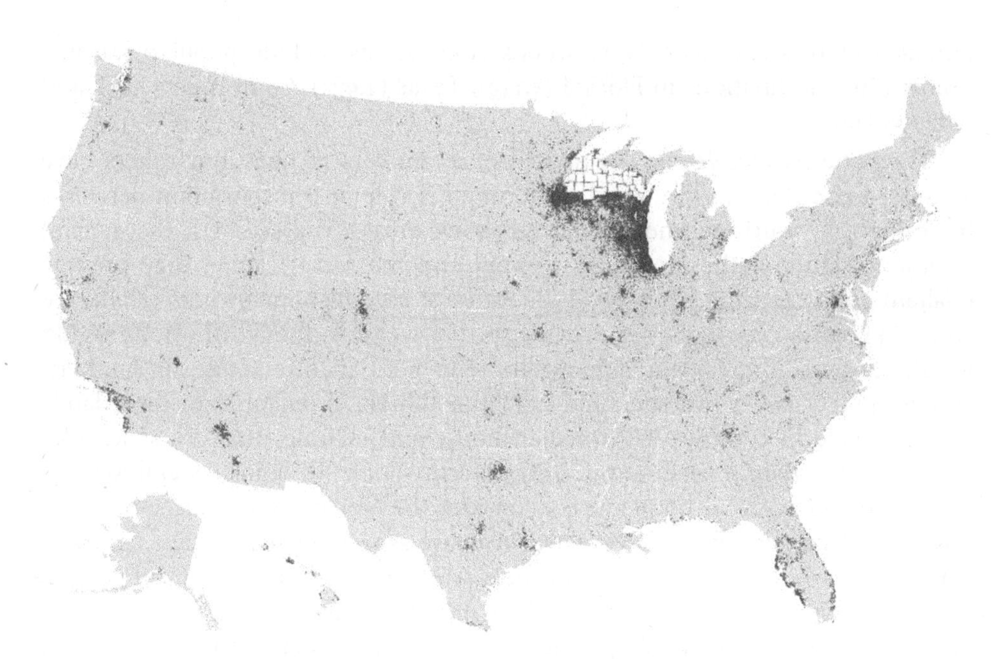

FIG. 2—Map of primary residences of nonresident northern Wisconsin landowners.

Nonresident owners of Northwoods property tend to live in urban and suburban regions of the Upper Midwest, and their properties, on average, are more valuable than the properties of permanent residents (Figure 2). Privately owned parcels with owner addresses in southern Wisconsin account for 19.5 percent of nonresident property owners (140,105 parcels), have an average parcel size of 18.4 acres, and an average assessed value of 95,408 USD, slightly higher than permanent residences. While 47,677 (34 percent) of these addresses are in the counties of the Milwaukee and Madison metropolitan regions, many more (59,893; 42.7 percent) are in counties immediately south of the study area. The proximity of the Northwoods to residents of these counties, which include the cities of Green Bay, Appleton, Wausau, and Eau Claire, is likely to explain the high rate of parcel ownership there.

Lastly, nonresident property owners with out-of-state addresses own 119,442 (16.6 percent) parcels, with an average parcel size of 16.2 acres and an average assessed value of 123,501 USD—more than 30,000 USD higher than parcels owned by local residents. Tax bills for these properties are sent to 50 states, 3 U. S. territories, and 25 countries. Minnesotans hold title to more of these parcels than any other state outside of Wisconsin. Collectively, they own 37,918 (31.9 percent of parcels with nonresident, out-of-state addresses), with 35,756 owned by people with addresses in the Twin Cities metropolitan area. Illinois residents own 24,536 (20.6 percent) of these parcels, and 21,575 of their addresses are located in the Chicago metropolitan area. Other states hosting noteworthy numbers of residents with Northwoods connections include popular seasonal "Snow Bird" destinations in Florida (4,149), Texas (2,977), California (2,593), and Arizona (1,635).

The patterns revealed in Figure 2 inspire answers to questions others have addressed elsewhere in the world, including the role of transportation networks in connecting tourism sending and receiving areas (Müller 2013), the general economic status of these seasonal visitors and the social capital they provide (Gallent 2014), and the heterolocal identities of second homeowners (Halfacree 2012). Limited-access highways provide speedy access to the Northwoods region for the 202,559 nonresident landowners who permanently reside in southern Wisconsin, eastern Minnesota, and northern Illinois. A majority of these landowners reside in counties with median household incomes that are at least 50 percent higher than most counties in the Northwoods (St. Louis Federal Reserve 2018). Although second homeowners consider the Northwoods a place of reprieve, heterolocal identities are reflected in many aspects of society. For example, many second homeowners invest in their seasonal communities by supporting local organizations. They also bring their metropolitan tastes to these communities, and the presence of shops and restaurants that cater to their preferences (and price points) reflect this. In the following section, however, we consider the interrelationship between second homes, property taxes, and school funding, an underexplored aspect of second home tourism mentioned by C. Michael Hall (2014, 121).

"Tax Exporting" as School Funding in the Northwoods

Current population trends for Wisconsin's rural communities are troublesome. In a recent study by Dan Barroilhet (2019), two-thirds of Wisconsin's rural counties, including 16 in the Northwoods study area, saw a drop in their total population between 2010 and 2018. David Egan-Robertson (2013) projects northern Wisconsin will witness some of the largest declines in county populations in upcoming decades, though these patterns throughout the area are not uniform. Bayfield, Ashland, Price, Rusk, Iron, and Florence counties are all projected to see between a 9 percent and 18 percent decline in total population by 2040. Conversely, Burnett, Washburn, Vilas, Douglas, Sawyer, Oneida, and Forest counties are expected to see between 5 and 15 percent growth. These counties are home to established tourism centers, including Hayward, Eagle River, and Minocqua, and concentrations of freshwater lakes with resorts and second homes. Counties losing populations tend to lack comparable tourism hubs and are also experiencing a loss of primary and secondary industries, which compounds population loss.

Though declining job opportunities are a major factor in explaining population loss, many rural counties of northern Wisconsin are also experiencing a two-fold process of out migration by younger persons and an aging residential population. The U.S. Census (2020) estimates that, between 2010 and 2018, Lincoln County experienced an exodus of 568 people and suffered 1,054 more deaths than births. Egan-Robertson (2013) reports that 11 of 14 Wisconsin counties with the highest proportion (20–30 percent) of people aged 65 or older in 2010 lived in Wisconsin's Northwoods. By 2040, the U.S. Census estimates that cohorts aged 65 and older will comprise more than 20 percent of the populations in all but two of Wisconsin's counties. Six of the eleven counties projected to have more than 35 percent of their population in these age cohorts (Bayfield, Iron, Rusk, Price, Vilas, and Florence) are in the Northwoods.

The realities of population decline for any given area are multifaceted. Though literature is plentiful regarding the impacts of population decline on local economies (Glaeser and others 1992; Puga 2010; Elshof and others 2014), Peter Matanle and Anthony Rausch (2011) observed that population decline also impacts the quality of life of effected communities. Perhaps hardest hit by these declines are public schools. Beyond the primary function of education, Karen Witten and others (2001) explained that schools provide secondary functions such as informational, material (public space), and social (friends or parents networks). Phillip Woods (2005) also identified schools as emotional and symbolic of a healthy and prosperous community. In many rural communities, the largest employer may be the public schools. Therefore, school closures can have devastating economic impacts on local communities. Sarah Kemp (2019) identified that rural school districts in Wisconsin have experienced a 7.4 percent decline in enrollment between 2006 and 2018. The decline in rural districts was not geographically uniform. School districts in the Northwoods of Wisconsin saw some of the largest decreases during this period. For example, the Florence, Goodman-Armstrong, and Drummond school districts experienced declines of more than 45 percent in their enrollment. Only handful of Northwoods school districts saw increases.

Additional burdens to Wisconsin's public education system were imposed in 2011, via Act 10, as Wisconsin legislators radically cut funds to public education and restricted teachers unions' ability to collectively bargain contracts. A 2018 UW-Madison study (Goff and others 2018) revealed that the starting salaries of incoming primary and secondary teachers declined dramatically and enrollment in education programs at four-year, public universities declined 35.4 percent between 2010 and 2016. Additionally, total teaching license endorsements declined by 17.2 percent. These realities are amplified in rural school districts. Closures of elementary schools and teacher shortages have become common in Wisconsin's Northwoods. A WXPR study (Meyer 2020) found that declining enrollments, declining tax bases, and funding difficulties forced several north-central Wisconsin school districts to close 27 elementary schools over the past 25 years. The lack of teachers also means schools must be creative in their approaches. A recent story by Rob Mentzer (2020) highlighted a blended online chemistry program and the cost-shared salary of its instructor between the

Phelps and Three Lakes school districts in Vilas and Oneida counties. In this example, Phelps students (the smaller of the two districts with a total of 87 K-12 students) will receive most of their education through video lectures and online assignments. Phelps School District Principal Jason Pertile said, "For smaller districts, the smaller you get, the more valuable those teachers become. With the teacher shortage, sometimes you just can't find somebody who wants to come live in northeastern Wisconsin" (Mentzer 2020).

These realities leave school districts few options. Combating funding shortages means closure, consolidation, or funding referenda. While only a handful of school districts are closed, a number have consolidated in some format. For those choosing some level of consolidation, this typically means consolidating activities like athletics, but similar approaches to the Phelps and Three Lakes examples are becoming more common. The most popular method to combat declining enrollments is to increase funding is by local referenda. According to the Wisconsin Department of Public Instruction (WPI 2020), the number of public referendums saw a dramatic increase after 2011. The peak occurred in 2014 at 202, up from 111 in 2009, but the total values of the referendums peaked in 2016. In this year, taxpayers across the state were asked to vote on nearly 2 USD billion in cumulative funding. In comparison, 2009 saw taxpayers deciding on a total 489 USD million in referenda funding.

Though referenda are a necessary part of public schools' survival, any increase in funding is passed on to the population through an increase in local property taxes. As populations decline in rural areas, remaining property owners can be further financially burdened. In 2016, had voters not approved a 675,000 USD referendum for the Lake Holcombe School District, leaders said they most likely would have had to close the school. Lake Holcombe School Board President, Corey Grape, explained, "A small rural school is really more than a school. We've been saying it's a lighthouse, it's the center of the community, it's the biggest industry Holcombe has" (Bringe 2016). More than 70 percent of the voters approved the referendum on 5 April 2016.

Second Home Ownership and Schools

Motivations for second home purchases are multifaceted. Raymond Burby and others (1972) found that friends and family were significant reasons why and where people purchase second homes. Bernadette Quinn (2004) cited a Canadian study that found 28 percent of second homeowners intended to retire in their second home and intend to reconnect with past experiences of a given place and escape the difficulties of daily life. The long history of tourism activities in the Northwoods was guided first by the rails and later highways. This allowed greater access by persons from all over Wisconsin, Minnesota, and Illinois. Early rail lines connected Eagle River and Hayward to Chicago, and improved access to Hayward from Minneapolis and St. Paul. Resorts dotted the shorelines of these communities providing visitors year-long recreational opportunities. This also provided visitors with lasting memories that were shared with friends, families, and neighbors, further illuminating these areas as tourist destinations.

Today, Hayward touts itself as the "Musky Capital of the World." To reinforce this narrative, travelers along Highway 27 are greeted by 143-foot-long, 45-foot-tall replica of a Muskellunge or "Musky," a freshwater trophy fish that can grow upwards of 48 inches in length. Eagle River boasts not only its lakes, but also its connections to winter. This city of 1,398 proclaims it is the "Snowmobile Capital of the World," as well as the "Hockey Capital of Wisconsin" (Eagle River Area Chamber of Commerce 2020). Both cities also claim many hotel rooms and cabins, campgrounds, evening boat cruises, ice cream shops, miniature golf, and biking trails. Though the winter months are less lucrative than the summer, the many lakes and bike trails give way to ice shanties for fishing, snowmobiles, and cross-country skiers.

Margaret Deery and others (2012) provide an extensive literature review on the social impacts of tourism. Specifically, they identify "value" variables that influence perceptions of tourism between visitors and residents. These include perceptions of community attachment and social, political, and environmental values (67). Early studies by George Doxey (1975) and Richard Butler (1980) theorize as communities become more imbedded in a tourist dependent economy, locals transform from an initial sense of euphoria to an eventual sense of antagonism toward tourists. Resulting perceptions are influenced by the duration of a community's ties to a strong tourist economy. R. Moisey and others (1996) found where tourism was a newer economic influence, initial reactions from residents was negative as tourists were perceived to be disruptive. Yet, in a place like Hawai`i, where tourism has long dominated the local economy, Liu and others (1987) identified locals who were unaware of the value of tourism. Faulkner and Tideswell's (1997) study of the impact on tourism on Australia's Gold Coast identified mixed results. Residents of the Gold Coast clearly identified the positive economic and social impacts of tourism to the area, but also identified frustrations with surges in traffic, noise, and increases in cost of living driven by an economy dependent on tourism.

Perhaps the best illustration of the love-hate relationship between locals and nonresidents in northern Wisconsin occurs every Labor Day in Minocqua, Wisconsin, where, for nearly 50 years, locals have sat in their lawn chairs along Highway 51 to wave and thank second homeowners and tourists as they depart for the summer. Shauna Whitman, a local bartender, explains: "Without the tourists there wouldn't be hospitals, there wouldn't be the schools that we have up here" (Porter 2019). Rich Stark, a landowner in the area, held a more nuanced view rationalizing the annual event stating: "[W]e're just telling them how great it was for them to share their time with us, but we're also happy to see they're going to leave us alone for a while" (Bowden 2018).

For communities to transform into tourist destinations, locals are "forced" to lose their identity as "owners of capital" through a process of "creative destruction." To accomplish the transformation, local officials must embrace tourism and eventually outside ownership of property. Throughout much of the Northwoods, as general demand for tourism increased, so too did the price of land, especially along the shorelines of lakes. Locals increasingly became unable to afford lakefront property. As wealth concentrated along the water, interest from outside populations increased as they were the only ones that could afford the land. This loss of local control increased the possibility of tension between local and nonlocal property owners. Michael Woolcock (1998) and Nick Gallent (2014) write that because many second homeowners are nonresidents, they may act on issues in their own interest because of the lack of connection to the local community. Nathan Anderson (2006) identified the most common tension between these two groups is often taxes. Specifically, Anderson identifies "tax exporting" as the ability of residents to impose a tax burden on nonresidents, thereby lowering the tax price of residents. This often puts locals and nonlocals at odds because nonlocals are often unable to vote on tax increases or in some states are required to pay higher tax rates than residents. Therefore, local officials are typically challenged to find a common ground between both sides. Increased tax rates may discourage potential buyers as well as put an extra burden on locals that typically have lower incomes.

As second home ownership increases, resulting tax revenues generally also increase. This benefits residents through improved local economies and infrastructure, lower tax rates, and better funded schools. In Wisconsin, local governments determine tax rates for public school funding. School district taxes are calculated by defining a mill rate that is then applied to the assessed value of a parcel of land. This is expressed as a cost-per-thousand dollars applied to a given property. For example, the 2018 mill rate for the Northland Pines School District in Eagle River was 5.88 per one thousand dollars. Therefore, a property worth 100,000 USD would pay 588 USD for school tax. Due to the reality of a "tax exporting" effect for nonlocal property owners in some Northwoods school districts, local officials must find a balancing act as not to create a perception of overburdening second homeowners by increasing the local tax rate. In their investigation of second home ownership in Michigan, Erik Johnson and Randall Walsh (2013) found that for a majority of second homeowners aged 50 or older, for every 100 USD increase in property taxes, there is a 0.73 percent increase in the probability that this same age group will move in the next two years. To avoid potential out migration of senior citizens, the Hayward School District passed a measure in 2017 offering a property tax exemption to district property owners 65 years or older (as of July 1 for any given year) that own and occupy a property as their primary place of residence. These individuals only pay 88 USD a year for twelve years (Hayward Unified School District 2017).

Referenda for school funding often create mixed reactions, as successful passage ultimately means increased tax rates for all property owners. They can be contentious for second homeowners because, as nonresidents, they are unable to participate in the final vote (Anderson 2006). A recent example of this controversy occurred in the Hayward School District in 2017. In the previous year, 54.4 percent of voters turned down a nearly 4.3 USD million referendum in 2016. Undeterred, district leaders brought the issue to vote a year later. In the weeks preceding the election, Letters to the Editor published in the *Sawyer County Record* highlighted the conflict nonresidents had with such laws. Brian Follett (2017), a nonresident homeowner from Woodbury, Minnesota: "Like many others, I own a vacation home in the Hayward area, pay taxes, and can't vote in the upcoming referendum" (2017, 4A). Another nonresident homeowner from St. Paul, Minnesota, echoed a similar response about paying higher property taxes (Anderson 2017). Locals who shared this sentiment expressed their frustrations with paying additional taxes in their own letters. The 2017 referendum failed by a 1 percent margin (Boettcher 2017). Interestingly, there is no record of contention by nonresident landowners in the *Vilas County Review* over the 13.8 USD million school referendum held by the Northland Pines School District in 2019. More than 65 percent of voters in there supported the property tax increase (Kimble 2019). In 2004, voters also approved a 28.5 USD million referendum to build a state-of-the-art high school for the district (see Figure 3).

Jason Giersch (2014) clearly indicated that the presence of second home ownership increases revenues for schools via property taxes. In comparing mill rates for school districts across the state of Wisconsin, both the Hayward School District and Northland Pines School District have some of the lowest mill rates in Wisconsin, at 6.3 and 5.88 respectively. In comparison, the highest value was the west central Wisconsin school district of Elmwood, at 14.99 (Anderson 2019).

FIG. 3—Photo of Northland Pines High School.

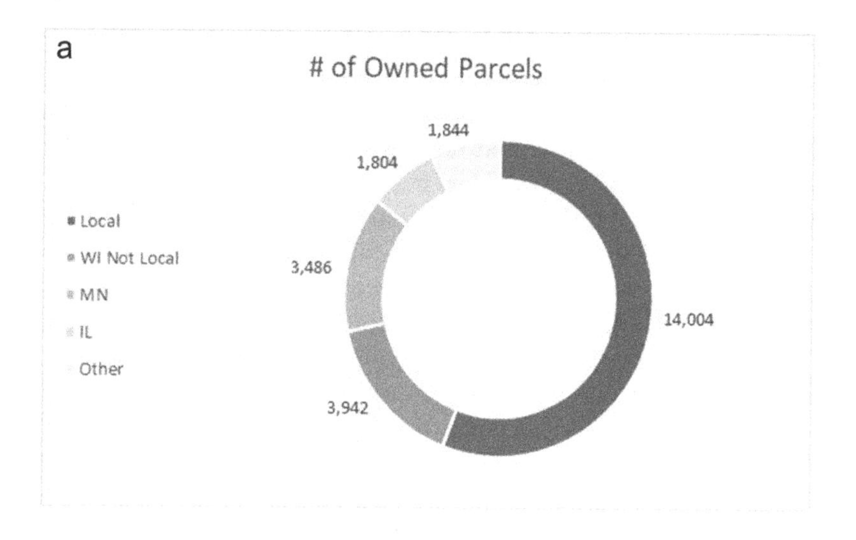

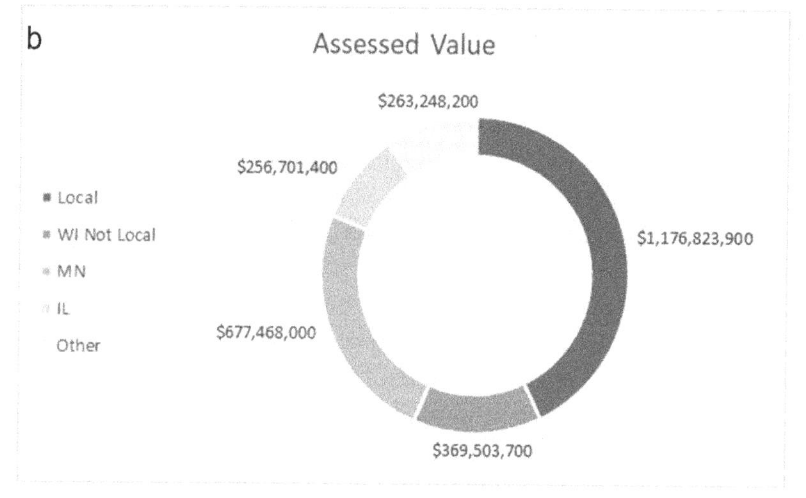

FIG. 4—Parcel and Assessed Value Information for Permanent Residents in the Hayward School District. Source: Wisconsin State Cartographer's Office. *Includes Figure 4a and 4b.

Mill rates can fluctuate with an addition of funding referendums for local school districts.

To gauge the impact of second home ownership in the areas throughout the Northwoods, a detailed analysis of parcel ownership was conducted for the each of the four school districts. Figures 4–7 identify the total number of parcels owned by locals, Wisconsin residents who are not local, Minnesota, Illinois, and "Other" areas. These figures also show the value of the parcels owned by the same groups of parcel owners. Though the Hayward School District is the largest in total area (358,590 acres), Northland Pines, the third largest among the four at 141,864 acres, holds the highest total assessed private property value, at more

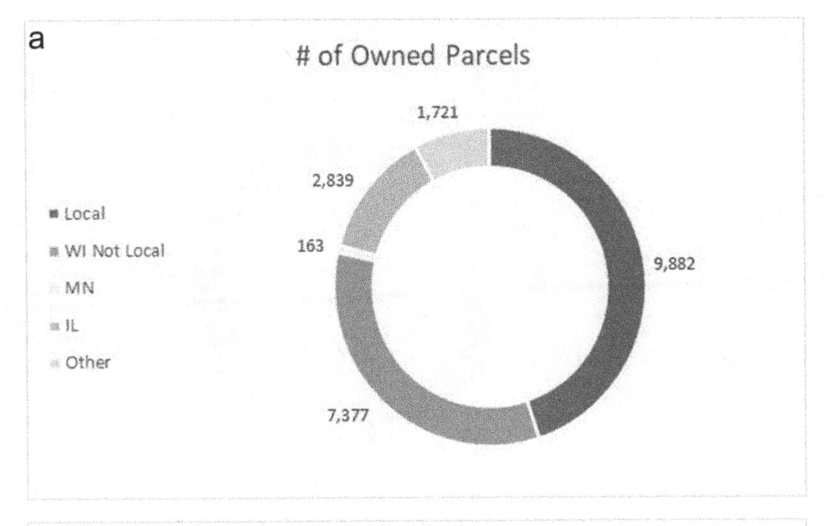

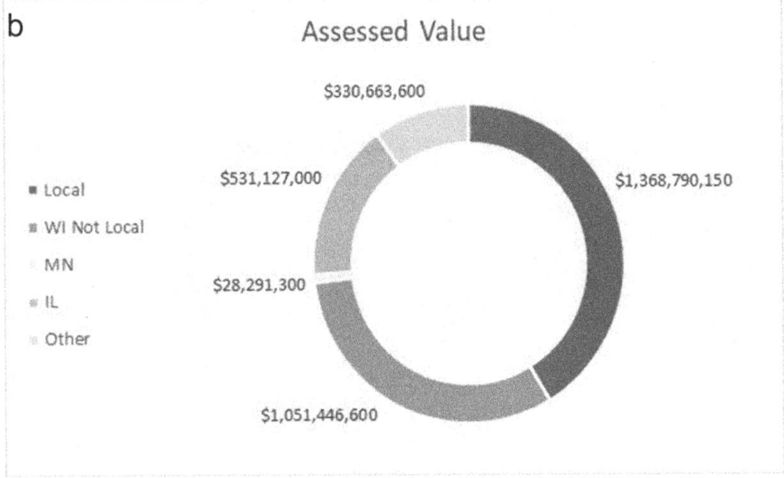

FIG. 5—Parcel and Assessed Value Information for Permanent Residents in the Northland Pines School District. Source: Wisconsin State Cartographer's Office. *Includes Figure 5a and 5b.

than 3 USD billion. Hayward comes in at a close second with 2.7 USD billion. The wealth of both districts is tied more to second homeowners than either Hurley or Grantsburg. Both Hayward and Northland Pines are owned by only 44 percent and 42 percent local residents, respectfully, while Hurley is nearly owned by 50 percent locals and Grantsburg by 66 percent of the local population. The resulting land values in each of the school districts is also reflected by total tax revenues. With nearly 20 USD million, the Northland Pines School District ranks in the top 40 statewide for tax revenue, the only rural school district in the state with this high a ranking. This revenue equates to 14,843.42 USD per student, followed closely by Hayward at 10,029.31 USD per student. The comparative lack of nonresidential lakeside property in Grantsburg and Hurley is

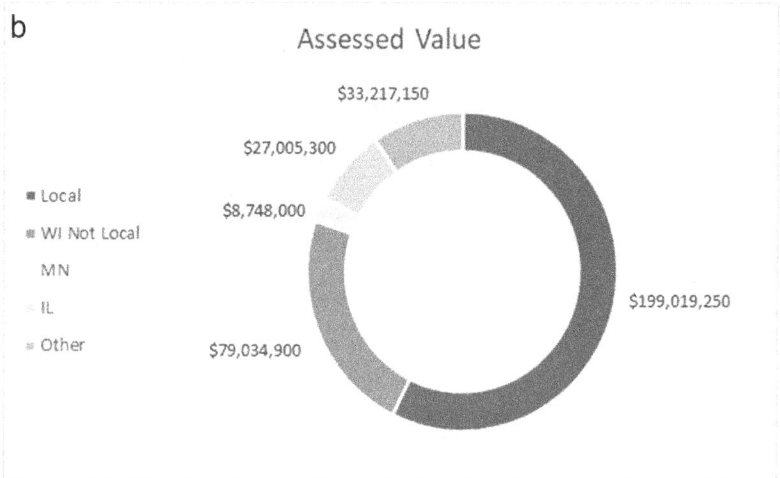

FIG. 6—Parcel and Assessed Value Information for Permanent Residents in the Hurley School District. Source: Wisconsin State Cartographer's Office. *Includes Figures 6a and 6b.

reflected by the higher levels of local ownership and lower values of the parcels. Tax revenue in these districts equates to only 6,090.83 USD per student in Hurley and 3,507.74 USD per student in Grantsburg.

The primary residences of parcel landowners in these school districts differs greatly between them. In all cases, the largest ownership of parcels is among local residents, followed by nonlocal Wisconsin residents, mainly from the Milwaukee and Madison metropolitan areas. Minnesota residents comprise the third highest number of landowners in the Grantsburg and Hayward school districts—not surprising given their proximity. Residents of Upper Michigan outnumber other out-of-state landowners in Hurley, and Illinois residents held the third highest

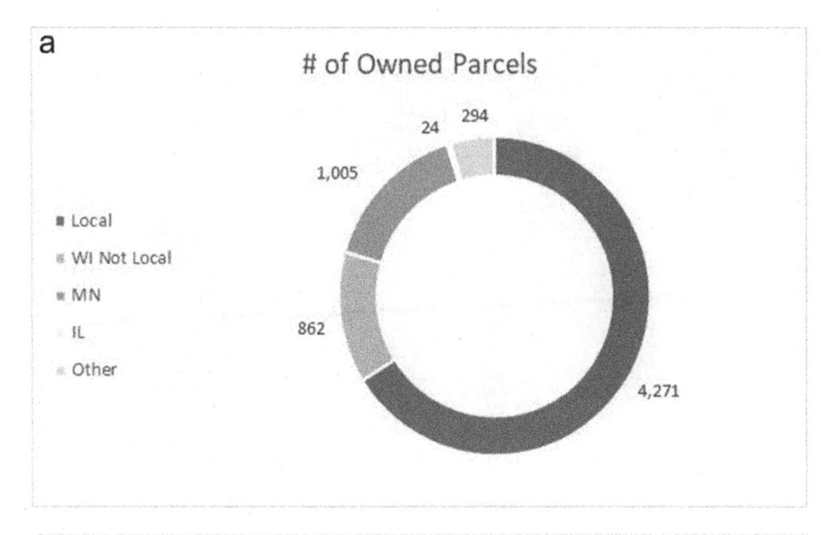

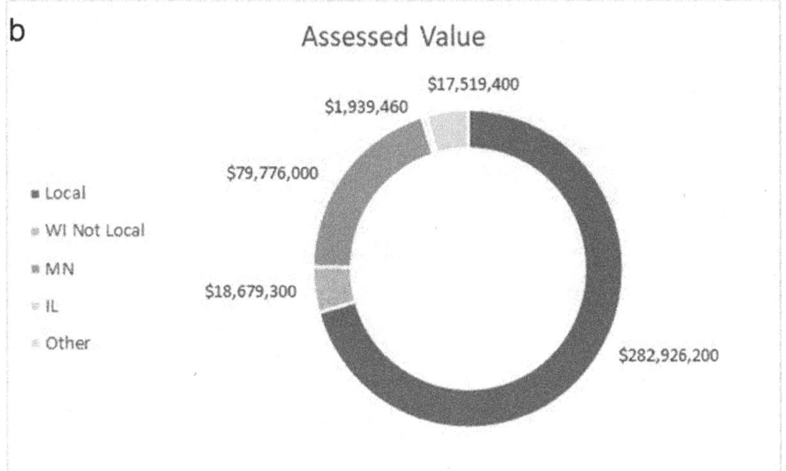

FIG. 7—Parcel and Assessed Value Information for Permanent Residents in the Grantsburg School District. Source: Wisconsin State Cartographer's Office. *Includes Figures 7a and 7b.

number of parcels in the Northland Pines School District. Data derived from the WSCO indicate that the property controlled by nonlocal Wisconsin residents and out-of-state residents is assessed higher than the property of permanent residents, particularly in the Hayward and Northland Pines school districts.

An in-depth spatial analysis (see Figures 8 and 9) of parcel ownership at the local level near the cities of Hayward and Eagle River clearly shows the ownership of lake parcels by nonlocal residents. These shoreline parcels also hold higher property values compared to parcels not along any body of water. This unmistakably indicates second home ownership attracts wealthier owners interested in utilizing the natural amenities waterfront properties provide. These realties also provide rural areas with well-developed tourism a clear economic

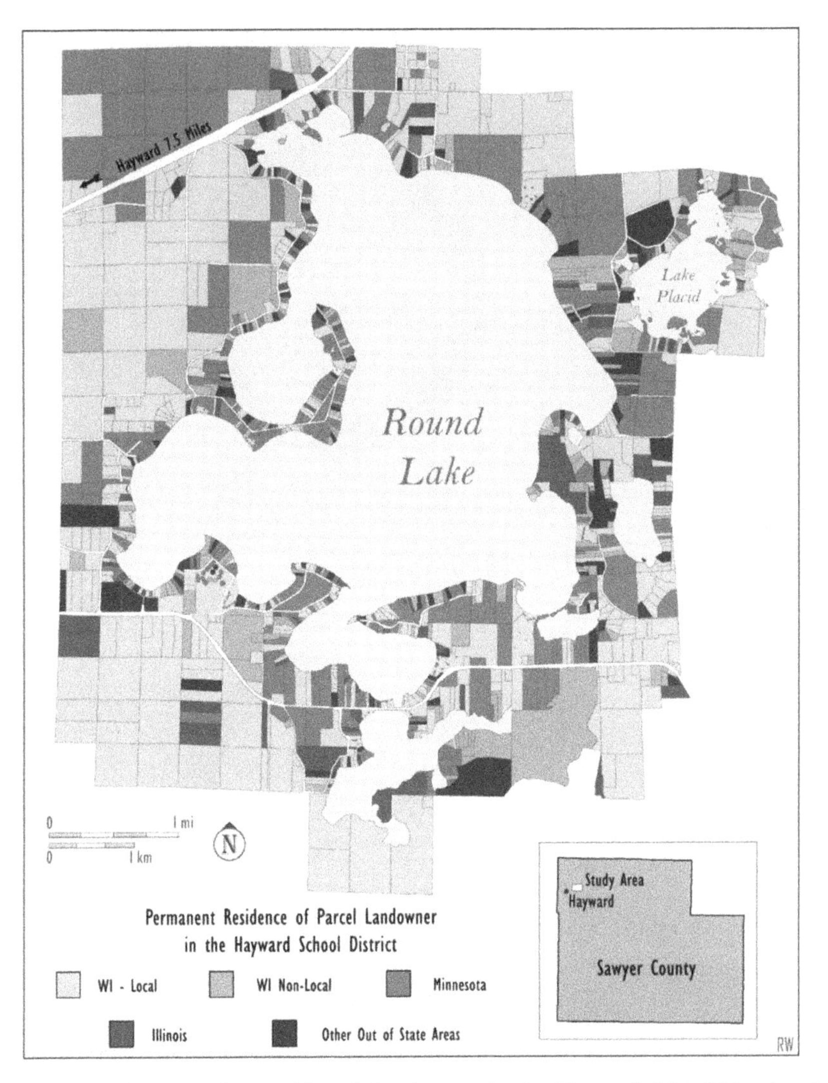

FIG. 8—Permanent Residence of Parcel Landowner in the Hayward School District. Source: Wisconsin State Cartographer's Office Statewide Parcel Data Set. Map produced by Ryan Weichelt.

advantage compared to other rural areas across the state. This symbiotic relationship allows all involved to keep tax rates lower, but due to the increased assessed value of land, especially along lakes, provides school districts economic stability through larger tax yields.

CONCLUSION

The impacts of tourism on the local landscape are multifaced. Interactions between tourists and full-time residents can and often does lead to some level of contention. While Doxey (1975) and Butler (1980) suggested locals'

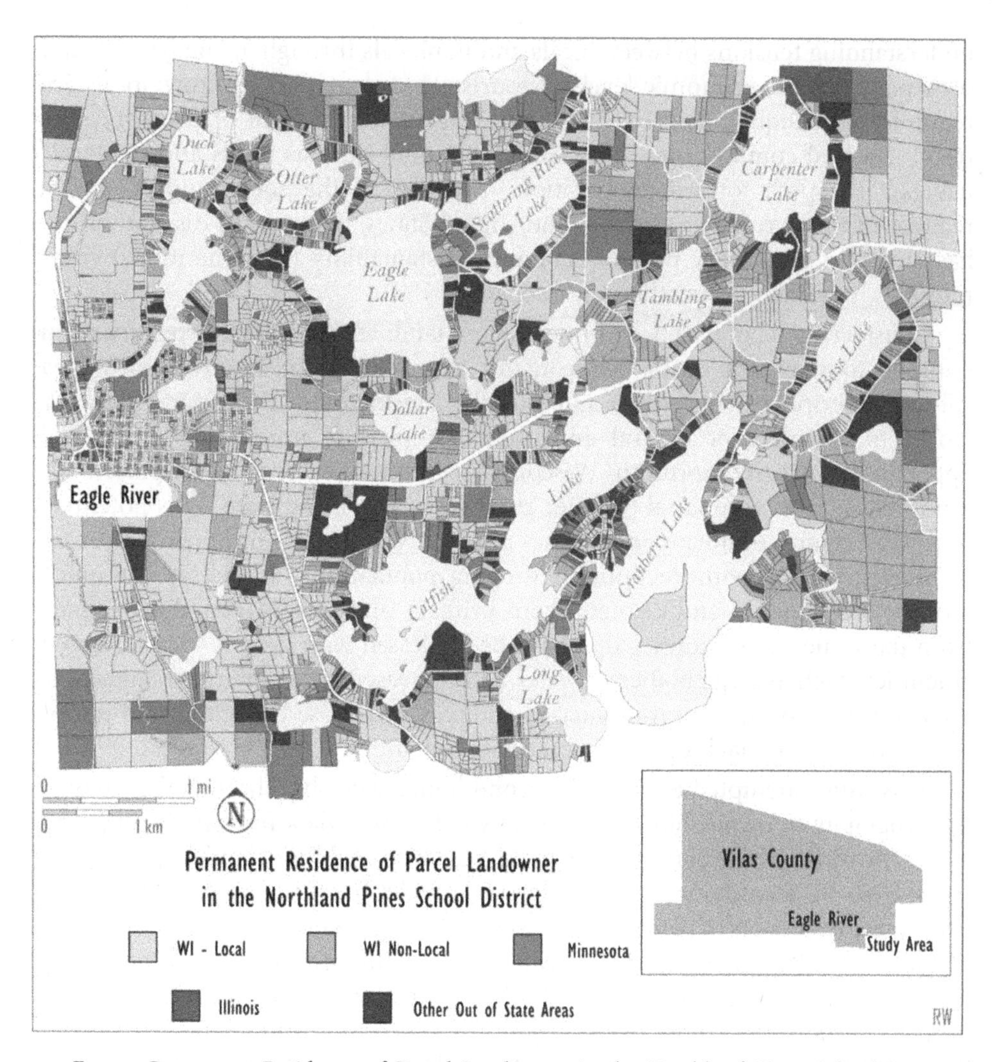

FIG. 9—Permanent Residence of Parcel Landowner in the Northland Pines School District. Source: Wisconsin State Cartographer's Office Statewide Parcel Data Set. Map produced by Ryan Weichelt.

perceptions on tourists typically ends in contention, Mosiey and others (1996) found differing reactions from locals with acceptance of the tourist economy while expressing only minor annoyances from visitors such as increased traffic and garbage. Juanita Liu and Turgut Var (1986) additionally found in their study of tourism in Hawai`i, most locals were unaware of the impacts tourists had on their everyday lives. Though these studies demonstrate that the perceived impacts of tourism vary place to place, they illustrate the ways that tourism transforms the physical and social landscape and that often impacts local residents more. Schumpeter's Theory of Creative Destruction provides a lens into understanding tensions between locals and nonlocals through an analysis of land ownership. As the economic levels of tourism-based economies grow, locals are required to relinquish control of tourism capital in the form of property. The result can lead communities to a state of disequilibrium as locals and nonlocals navigate the path toward cooperation. Yet, due to the social and economic realities of many rural areas across the United States, this state of disequilibrium has created a temporary path of economic sustainability for some tourist communities, especially in northern Wisconsin.

Considering its prevalence of nonresident landownership and dependence on outdoor tourism amenities, we contend that Wisconsin's Northwoods is a "pleasure periphery" for urban and suburban residents of the Upper Midwest. Home to a variety of natural amenities that provide numerous recreational opportunities, many northern Wisconsin communities are better poised for economic success compared to other rural areas across the state. This potential, however, is not uniform across the region. Nearly all northern Wisconsin counties are faced with declining and aging populations, as well as continued economic decline, as employment from primary and secondary industries fade from the scene. Communities that have been blessed with greater tourism-based amenities, such as large bodies of water and increased accessibility via highways, are better poised for economic success compared to those that either did not or could not due to a lack of popular tourism amenities.

This study attempted to quantify second homeownership through the study of the geography of ownership of the Northwoods. Using data from individual land parcels provided by the State of Wisconsin, a detailed analysis of landownership was conducted. By identifying the tax address provided in this robust database, the permanent residences of parcel owners were able to be identified. The first approach of this study clearly identified the spatial patterns of lands ownership of second homes in Wisconsin's Northwoods. While these patterns noticeably indicated that property owners were more likely to be from Wisconsin, Minnesota, and Illinois, the use of the tax address allowed a better understanding of where second homeowners make their permanent residence. While every parcel and their owners have a story to tell, it is evident that parcel ownership is driven by suburban and urban populations across Wisconsin's east shore through Madison, as well as persons from the Minneapolis-St. Paul metropolitan area and Chicago and its environs. Further comparisons between the value of locally owned and nonresident parcels identified a significant disparity between the value of those owned by residents and those owned by out-of-state residents.

The second approach to this study focused on the potential economic impacts of second home ownership on specific school districts. Given the information provided by the State of Wisconsin, it was possible to determine the impact of both residents and nonresidents on property tax collections. Through the analysis of four distinct school districts, it was clear that the tourist-centered communities of the

Hayward and Eagle River greatly benefited from the investment of second homeowners, particularly when contrasted to Hurley and Grantsburg districts that lacked the same amount of tourist-related amenities and nonresident property owners. Figures 7 and 8 provided a clearer scope of second homeownership around a few of the many lakes in the Hayward and Northland Pines school districts. Lakefront property had a much higher value than off-water property, and as the figures show, these properties were more likely owned by nonlocal residents. These patterns illustrate the overall increased land values in the two districts that translated to much higher tax revenues. With an enrollment of only 1,289 students, Northland Pines generated nearly 20 USD million in local school tax revenue, equating to over 14,000 USD per student. The impact of second home ownership is obvious when compared to the smaller and less tourist-dominated economy of the Hurley school district, equating to roughly 6,000 USD for each of the 562 students. A secondary benefit for Hayward and Northland Pines landowners is a much lower tax rate than other areas.

This study offers a distinct analysis of the geography of second homeownership, as the incorporation of individual parcel data and the tax address provides a greater understanding of who controls the most valuable land in the Northwoods and where those landowners reside. Furthermore, this analysis provided an opportunity to gauge the influence of local and nonlocal landowners on the local economy and on the local school district. Considering the demographic and economic trajectory of Wisconsin's Northwoods, places in a position to accommodate second homeowners are better situated to thrive than places lacking natural amenities that tourists desire. Yet, economic success does not come without strings attached, as nonresident landowners have obvious influence over economic activities and involvement in local governmental decisions in Hayward and Eagle River. Though lower tax rates are generally things most people will not complain about, Northwoods communities find they are necessary to attract new investment and keep nonresident landowners happy. Unfortunately, the high price of land excludes many locals from purchasing the most desirable property, leaving the economic future of these areas in the hands of nonlocals.

The examples in this study provide a glimpse of the realities faced by local officials in balancing the needs of all landowners. The availability of such a robust dataset provided by the State of Wisconsin provides a unique window into the patterns of landownership, yet it lacks the ability to understand the impacts on individuals. While we are not ignoring these impacts, they are beyond the scope of this study. Future studies focusing on second home ownership incorporating not only secondary data such as homeowner's address or property value, but rather a mixed methods approach, could solidify understandings of frustration across communities with varying tourist economies. As this study illustrates, second home ownership in amenity-rich areas of Wisconsin's Northwoods sustains local economies and school systems. To do so, however, communities concede access their most valuable amenities to nonresidents and cater to many of their preferences.

Orcid

Ryan Douglas Weichelt http://orcid.org/0000-0001-6515-0294

Ezra Zeitler http://orcid.org/0000-0003-3692-3010

References

Adams, B. 2014 24 May. Up North, Not on a Map but a Prolific Wisconsin Place. *Wisconsin State Journal*. https://host.madison.com/wsj/travel/local/up-north-not-on-a-map-but-a-prolific-wisconsin/article_cbff5360-16dd-59a3-bd4e-63b0de80c105.html

Alanen, A. 1997. Homes on the Range: Settling the Penokee-Gogebic Iron Ore District of Northern Wisconsin and Michigan. In *Wisconsin Land and Life*. edited by R. C. Ostergren and T. R. Vale. 241–262. Madison: University of Wisconsin Press.

Anderson, N. B. 2006. Beggar Thy Neighbor? Property Taxation of Vacation Homes. *National Tax Journal* 59(4): 757–780.

Anderson, S. 2017. When No Doesn't Seem to Mean No. *Sawyer County Record*, 15 March, 4A.

———. 2019. Every 2018–19 Wisconsin School District Tax Rate Ranked. *Patch*, 18 April. https://patch.com/wisconsin/mountpleasant/every-2018-19-wisconsin-school-district-tax-rate-ranked

Barroilhet, D. 2019. Wisconsin Demography: Present Profile and Future Trajectory, 19 August. https://www.wicounties.org/uploads/EventMaterials/wi-counties-association.pdf

Bawden, T. 1997. The Northwoods: Back to Nature? In *Wisconsin Land and Life*. edited byR. C. Ostergren and T. R. Vale. 450–469. Madison: University of Wisconsin Press.

Boettcher, R. 2017. *Concerns Raised Post-Referendum*. Sawyer County Record, 19 April, 4A.

Bowden, B. 2018. "Wave Parties" Bid Goodbye to Summer, Tourists. *Wisconsin Life*, 14 September. https://www.wisconsinlife.org/story/wave-parties-bid-goodbye-to-summer-tourists/

Bringe, J. 2016. Referendum Ensures Lake Holcombe School District Remains Open. *WEAU*, 6 April. https://www.weau.com/home/headlines/Referendum-ensures-Lake-Holcombe-School-District-remains-open-374827141.html

Burby, R., T. Donnelly, and S. F. Weiss. 1972. Vacation Home Location: A Model for Simulating the Residential Development of Rural Recreation Areas. *Regional Studies* 6(4): 421–439.

Butler, R. 1980. The Concept of the Tourist Area Cycle of Evolution: Implications for the Management of Resources. *Canadian Geographer* 24: 5–12.

Deery, M., L. Jago, and L. Fredline. 2012. Rethinking Social Impacts of Tourism Research: A New Research Agenda. *Tourism Management* 33(1): 64–73.

Doxey, G. V. 1975. A Causation Theory of Visitor Resident Irritants: Methodology and Research Inferences. Paper presented at the Travel and Tourism Research Association sixth annual conference, San Diego, CA.

Eagle River Area Chamber of Commerce. 2020. The Making of a Community, 3 February. https://eagleriver.org/

Egan-Robertson, D. 2013. Projections for the State, Its Counties and Municipalities, 2010 – 2040, December. https://fyi.extension.wisc.edu/agingfriendlycommunities/files/2014/08/2013-Egan-Robertson_Wisconsin-Future-Population.pdf

Elshof, H., L. V. Wissen, and C. H. Mulder. 2014. The Self-reinforcing Effects of Population Decline: An Analysis of Differences in Moving Behaviour between Rural Neighborhoods with Declining and Stable Populations. *Journal of Rural Studies* 36: 285–299.

Faulkner, B., and C. Tideswell. 1997. A Framework for Monitoring Community Impacts of Tourism. *Journal of Sustainable Tourism* 5(1): 3–28.

Follett, B. 2017. Writer Gives Perspective on Hayward Referendum. *Sawyer County Record*, 22 March, 4A.

Gallent, N. 2014. The Social Value of Second Homes in Rural Communities. *Housing, Theory and Society* 31(2): 174–191.

Giersch, J. 2014. Effects of Vacation Properties on Local Education Budgets. *Cogent Economics & Finance* 2(1): 941890.
Glaeser, E. L., H. D. Kallal, J. A. Scheinkman, and A. Schleifer. 1992. Growth in Cities. *Journal of Political Economy* 100: 1126–1152.
Goff, P., B. Carl, and M. Yang. 2018. Working Paper. *Supply and Demand for Public School Teachers in Wisconsin.* Madison: Wisconsin Center for Education Research.
Gough, R. 1997. *Farming the Cutover: A Social History of Northern Wisconsin, 1900–1940.* Lawrence: University of Kansas Press.
Hagen, J. 2006. *Preservation, Tourism, and Nationalism: The Jewel of the German Past.* Hampshire, U.K: Ashgate.
Halfacree, K. 2012. Heterolocal Identities? Counter-Urbanisation, Second Homes, and Rural Consumption in the Era of Mobilities. *Population, Space and Place* 18(2): 209–224.
Hall, C. M. 2014. *Tourism and Social Marketing.* London: Routledge.
Hart, J. F. 1984. Resort Areas in Wisconsin. *Geographical Review* 74(2): 192–217.
Harvey, D. 1985. *The Urbanization of Capital.* Oxford, U.K: B. Blackwell.
———. 1988. The Geographical and Geopolitcal Consequences of the Transition from Fordist to Flexible Accumulation. In *America's New Market Geographies: Nation, Region and Metropolis.* edited byG. Sternlieb and J. W. Huges. 101–136. Rutgers, N.J: Center for Urban Policy Research.
Johnson, E., and R. Walsh. 2013. The Effect of Property Taxes on Vacation Home Growth Rates: Evidence from Michigan. *Regional Science and Urban Economics* 43(5): 740–750.
Jones, M. 2018 13 June. Cabin Life up North Is a Summer Generational Ritual in Wisconsin. *Milwaukee Journal-Sentinel.* https://www.jsonline.com/story/news/local/wisconsin/2018/06/13/opening-summer-cabins-wisconsins-northwoods-spans-generations/667658002/
Kates, J. 2001. *Planning a Wilderness: Regenerating the Great Lakes Cutover Region.* Minneapolis: University of Minnesota Press.
Kemp, S. 2019. Why More Wisconsin Schools Are Enrolling Fewer Students. *WisContext,* 9 September. https://www.wiscontext.org/why-more-wisconsin-schools-are-enrolling-fewer-students.
Kimble, L. 2019. WJFW - Northland Pines School District Asking Voters for $4.6 Million Per-year Referendum in February. *WJFW,* 19 February. https://www.wjfw.com/storydetails/20190103163352/northland_pines_school_district_asking_voters_for_46_million_peryear_referendum_in_february
Lewis, C. 2017 27 July. Wisconsin's Northwoods Abound in Lakes, Trees and that Special 'Up North' Feeling. *Milwaukee Journal-Sentinel.* https://www.jsonline.com/story/travel/wisconsin/2017/07/27/wisconsins-northwoods-abound-lakes-trees-and-special-up-north-feeling/507899001/
Liesch, M. 2006. *Ironwood, Hurley, and the Gogebic Range.* Charleston, S.C: Arcadia Publishing.
Lindmark, R. 1971. Second Homes in Northwestern Wisconsin: A Study of the Owners and Their Use Patterns and Characteristics of the Second Home Structure. Electronic Thesis or Dissertation. Ohio State University. https://etd.ohiolink.edu/
Liu, J. C., and T. Var. 1986. Resident Attitudes toward Tourism Impacts in Hawaii. *Annals of Tourism Research* 13: 193–214.
Liu, J. C., P. J. Sheldon, and T. Var. 1987. Resident Perception of the Environmental Impacts of Tourism. *Annals of Tourism Research* 14: 17–37.
Loew, P. 2013. *Indian Nations of Wisconsin: Histories of Endurance and Renewal.* 2nd. Madison: University of Wisconsin Press.
Marcouiller, D. W., G. P. Green, S. C. Deller, and N. R. Sumathi. 1996. *Recreational Homes and Regional Development: A Case Study from the Upper Great Lake States.* Madison: University of Wisconsin-Madison Extension.
Matanle, P., and A. S. Rausch. 2011. *Japans Shrinking Regions in the 21st Century: Contemporary Responses to Depopulation and Socioeconomic Decline.* Amherst, N.Y: Cambria Press.
Mentzer, R. 2020. These Rural Northwoods Schools Have a Teacher-Sharing Agreement. *WPR,* 27 January. https://www.wpr.org/these-rural-northwoods-schools-have-teacher-sharing-agreement

Meyer, B. 2020. Disagreement over Merrill School Symbolizes Rural Elementary Closure Trend in Northwoods. WXPR, 28 January. https://www.wxpr.org/post/disagreement-over-merrill-school-symbolizes-rural-elementary-closure-trend-northwoods#stream/0.

Mitchell, C. 1998. Entrepreneurialism, Commodification and Creative Destruction: A Model of Post-modern Community Development. *Journal of Rural Studies* 14(4): 273–286.

Moisey, R. N., N. P. Nickerson, and S. F. McCool. 1996. Responding to Changing Resident Attitudes toward Tourism: Policy Implications for Strategic Planning. Paper Presented at the 27th Annual Conference of the Travel and Tourism Research Association, Las Vegas, Nev.

Moran, D. 2018. Getting (Mostly) off the Grid Deep in Wisconsin's Northwoods. *Lake County News-Sun*, 15 August. https://www.chicagotribune.com/suburbs/lake-county-news-sun/ct-lns-moran-northwoods-wisconsin-st-0811-story.html

Müller, D. K. 2005. Second Home Tourism in the Swedish Mountain Range. In *Nature Based Tourism in Peripheral Areas: Development or Disaster?* edited byC. M. Hall and S. W. Boyd. 133–148. Clevedon, U.K: Channel View Publications.

———. 2013. Second Homes: Curse or Blessing? A Review 36 Years Later. *Scandinavian Journal of Hospitality and Tourism* 13(4): 353–369.

Nelson, J. B. 2017 29 June. Where Does "Up North" Begin? Wisconsinites Can't Agree. *Milwaukee Journal Sentinel.* http://www.jsonline.com/story/news/local/wisconsin/2017/06/29/where-does-up-north-begin-wisconsinites-cant-agree/438572001

Porter, S. 2019. Minocqua Residents Gather to Wave Goodbye to Tourists. *WSAW*, 2 September. https://www.wsaw.com/content/news/Minocqua-559197271.html

Puga, D. 2010. The Magnitude and Causes of Agglomeration Economies. *Journal of Regional Science* 50(1): 203–219.

Quinn, B. 2004. Dwelling through Multiple Places: A Case Study of Second Home Ownership in Ireland. In *Tourism, Mobility and Second Homes.* edited byC. M. Hall and D. K. Müller. 113–130. Clevedon, U.K: Channel View Publications.

Rohe, R. 1997. Lumbering Wisconsin's Northern Urban Frontier. In *Wisconsin Land and Life.* edited byR. C. Ostergren and T. R. Vale. 221–240. Madison: University of Wisconsin Press.

Satz, R. N. 1991. *Chippewa Treaty Rights: The Reserved Rights of Wisconsin's Chippewa Indians in Historical Perspective.* Madison: Wisconsin Academy of Sciences, Arts and Letters.

Schumpeter, J. 1942. *Capitalism, Socialism and Democracy.* New York: Harper & Bros.

See, C. 2017. Version 3 Statewide Parcel Database Completed. *Wisconsin State Cartographer's Office Online*, 15 July. http://www.sco.wisc.edu/find-data/parcels.html

Senior Citizen Tax Exemption. 2017. Hayward Unified School District, 2 May. https://www.husd.us/seniorcitizentaxexemption

Shapiro, A. 2013. *The Lure of the North Woods: Cultivating Tourism in the Upper Midwest.* Minneapolis: University of Minnesota Press.

Solomon, A. 2004. The Lure of the North Woods. *Chicago Tribune*, 25 July. https://www.chicagotribune.com/travel/chi-0407240222jul25-story.html

St. Louis Federal Reserve. 2018. Estimate of Median Household Income. *FRED Economic Research - Results.* https://fred.stlouisfed.org/searchresults/?nasw=0&st=medianpercent20householdpercent20income&t=county&ob=sr&od=desc&types=gen;geot

Stedman, R. C. 2003. Is It Really Just a Social Construction?: The Contribution of the Physical Environment to Sense of Place. *Society and Natural Resources: An International Journal* 16(8): 671–685.

———. 2006. Understanding Place Attachment Among Second Home Owners. *American Behavioral Scientist* 50(2): 187–205.

Turner, L., and J. Ash. 1975. *The Golden Hordes: International Tourism and the Pleasure Periphery.* London: Constable.

United States Census—Data Access and Dissemination Systems (DADS). 2020. American FactFinder—Results. https://factfinder.census.gov/faces/tableservices/jsf/pages/productview.xhtml?pid=PEP_2018_PEPTCOMP&prodType=table.

Vanderwerf, J. L. 2008. Creative Destruction and Rural Tourism Planning: The Case of Creemore. Masters Thesis, Waterloo University.

Weichelt, K. 2016. *A Historical Geography of the Paper Industry in the Wisconsin River Valley.* PhD Dissertation, University of Kansas. https://kuscholarworks.ku.edu/handle/1808/21911

Wisconsin Department of Administration. 2013. Act 20 and the Wisconsin Land Information Program, 17 July. https://doa.wi.gov/DIR/Act-20-and-WLIP-2013-07-17.pdf

Wisconsin Department of Natural Resources. 2020. Snowmobiling in Wisconsin. https://dnr.wisconsin.gov/topic/Snowmobile

Wisconsin Department of Tourism. 2020. County Total Economic Impact, 1 May. http://industry.travelwisconsin.com/research/economic-impact

WISE Data Elements. 2020 Wisconsin Department of Public Instruction. https://dpi.wi.gov/wise/data-elements.

Witten, K., T. Mccreanor, R. Kearns, and L. Ramasubramanian. 2001. The Impacts of a School Closure on Neighbourhood Social Cohesion: Narratives from Invercargill, New Zealand. *Health & Place* 7(4): 307–317.

Woods, P. 2005. *Democratic Leadership in Education*. London: Paul Chapman Publishing.

Woolcock, M. 1998. Social Capital and Economic Development: Towards a Theoretical Synthesis and Policy Framework. *Theory and Society* 27(27): 151–208.

WPI Custom Referenda Reports. 2020. Custom Referenda Reporting. https://sfs.dpi.wi.gov/Referenda/CustomReporting.aspx.

RE-TURNING INWARDS OR OPENING TO THE WORLD? LAND USE TRANSITIONS ON AUSTRALIA'S WESTERN COAST

ROY JONES, TOD JONES and COLIN INGRAM

ABSTRACT. Prior to European Settlement in 1829, the Western Australian coast to the north of Perth, the state capital, had long been occupied by the Yued Nyungar Aboriginal group. However, much of this land had limited agricultural potential and, following Aboriginal dispossession, it remained as largely unoccupied Crown (public) Land for about a century. From the 1920s, farmers, crayfishers and Perth residents began to establish campsites and shacks for temporary use. However, since the 1960s, pressure has been growing: to develop better access routes and more formal (and legal) coastal/recreational settlements; to offer greater statutory protection to the natural coastal environment; and to acknowledge Aboriginal rights over some areas of Crown Land. This paper analyses the land use transitions experienced in this coastal area, with particular reference to the growing and diversifying external pressures that are being applied to this formerly isolated and currently vulnerable locality.

Our paper analyses land use transitions on a 100 kilometer stretch of the Western Australian coast from before the onset of European settlement in the nineteenth century to the present day. Its temporal coverage therefore extends beyond the current era of globalization and economic nationalism, which is one component of this issue's key organizing theme. Rather, it addresses the issues of rural sustainability and vitality through the lens of the ongoing tensions between local desires for stability—which are themselves diverse and conflicting—in this decreasingly isolated locality and the varied external pressures for change. While some of these external pressures relate to economic development, they also include environmental conservation initiatives and the acknowledgment and formalization of Indigenous land ownership and management rights, all of which are now global phenomena. The spatial focus of this study is the coastal zone of the Shire (i.e. the local government area) of Dandaragan which is approximately 100–200 kilometers north of the Western Australian state capital of Perth (Figure 1). Although this region is relatively close to a colonial city which was first settled by Europeans in 1829, and which has since grown into a metropolitan area with a population of around 2,000,000, Dandaragan's coastal zone had little or no agricultural potential (McArthur and Bettenay 1960), and a sealed road through the area was only constructed in 2010. It therefore remained isolated for much of Western Australia's almost 200-year post invasion history and,

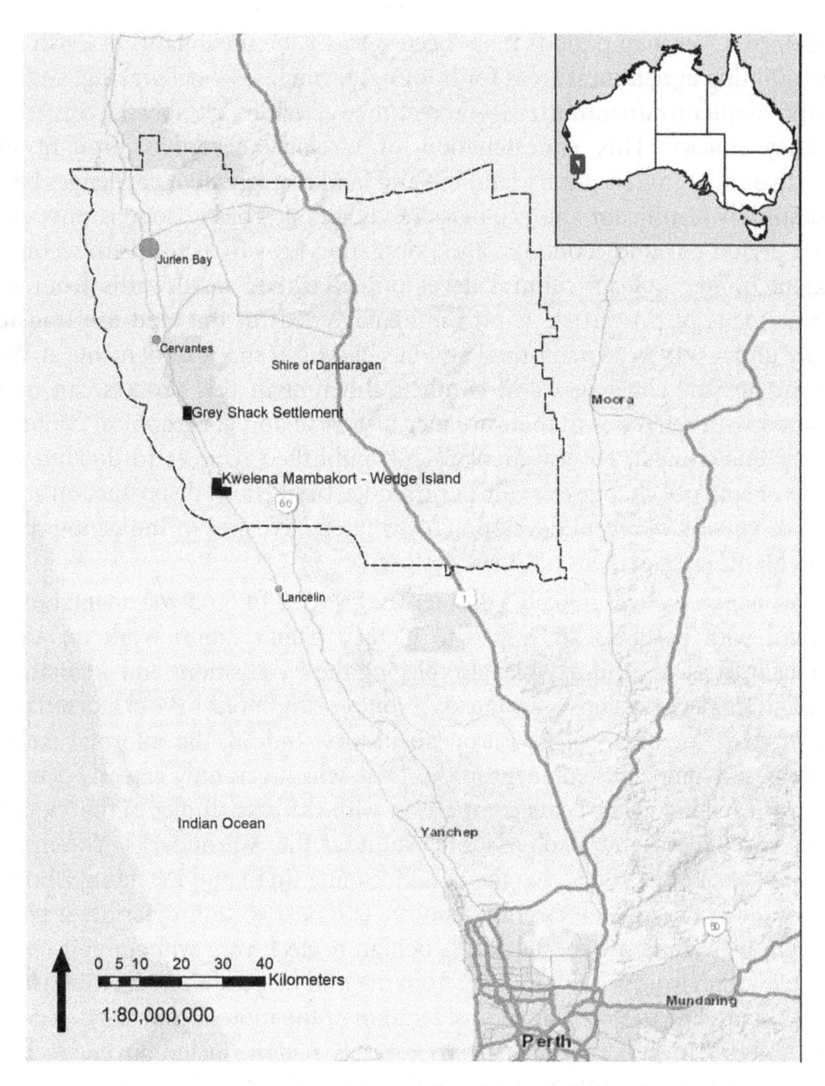

FIG. 1—Location map of the Dandaragan coastal zone (Tod Jones).

for much of this period, it played little or no part in in the state's development trajectory. Insofar as this was based on economic nationalism, the state's development has been tied to imperial and, subsequently, global markets for agricultural products and mineral resources, and was heavily dependent on international migration and capital (Crowley 1970; Stannage 1981).

As a result of its isolation and perceived lack of economic potential, the land use transition of this coastal area from Indigenous dispossession in the early nineteenth century to closer incorporation into the state's, and the world's, economy and society in recent decades has been slow and sporadic and has included numerous idiosyncrasies. Most of the land remains in government ownership and is largely undeveloped. Over long periods it has been used by the inhabitants of Perth and of the surrounding agricultural areas for hunting, fishing, seasonal grazing, and informal—and frequently unauthorized—recreational activities, notably the construction of holiday shacks. This concatenation of circumstances has brought about a distinctive set of contemporary land use and land management challenges because this botanically significant and ecologically vulnerable coastal zone is now experiencing a period of rapid economic and population growth as tourism, second and retirement home, and agricultural development extend northwards from Perth. A narrow focus of this article is on the management of the land use transitions currently under way in two informal and illegal coastal shack settlements at Wedge and Grey, but the challenges and conflicts inherent in this process can only be understood with reference to their broader historical and geographical contexts. In providing this context, we use archival and published sources to document two centuries of land use change over the period since Indigenous dispossession and link this to a discussion of recent developments whereby a degree of Indigenous control over local land management has been restored.

In this paper, we will initially consider the process of land use adaptations and transitions, with particular reference to Craig Colten's (2019) work on adaptive transitions in coastal settings, while also placing these transitions and adaptations in frontier and modes of occupance contexts. Frontier conditions have characterized this stretch of coast for much of its European history. Indeed, the informal nature of Indigenous and 'pioneer' settlement in this area, which was only recently deemed to be of environmental value, bears comparison with the case studies of the "Squatters, Poachers and Thieves" in Jacoby's (2014) study of the Adirondacks, Yosemite and Grand Canyon. We then describe the period leading up to and including Aboriginal dispossession in the mid nineteenth century. The next section describes a phase of what might be termed the coastal zone's benign neglect from without and opportunistic exploitation from within, roughly from the mid nineteenth to the mid twentieth century. This will be followed by a consideration of the more recent stage of growing external interest in and official concern for this region, including the increasing influence of both native title and recognition of Indigenous interests in land. This sequence of the local development of informal settlements followed by external attempts to formalize, regulate or even supplant these settlements again has parallels across Australia (P.I.R.G 1977; Drew 1994) and beyond (Hardy and Ward 1984; Selwood 2006). In conclusion, we assess the prospects for both the Indigenous and the shack settlement populations as components of a local rural system which is at least planned to become both sustainable and vital as this area develops novel and stronger connections to the wider world.

Our information sources encompass periodic visits to the shack settlements over more than a decade, numerous informal discussions with 'shackie' settlers, unstructured interviews with leaders of the Wedge and Grey community associations, participant observation at meetings over the future of these settlements, the collection and analysis of 'ephemera', from posters and leaflets to flyers and calendars, produced by the associations and qualified access to the files and reports of relevant state government departments.

Adaptations, Transitions, Frontiers and Modes of Occupance

In his study of the adaptive transition of the Louisiana coastline over a comparable period, Colten (2019) juxtaposes the primarily ecological concept of adaptation (Darwin 1859) and the socioeconomic concept of historical transitions (Wallerstein 1976), which he sees as being in an hierarchical relationship with each other. Adaptations are "human actions taken in an effort to perpetuate a society, even if modified in some way". Such adaptations "do not fundamentally alter society" and, while "they often take place over years or decades, (they) are often local or regional in scale" (Colten 2019, 417). Transitions occur as a result of "multiple adaptations, deliberate or ad hoc, coordinated or uncoordinated" and do not "presuppose human actions are reducible to systematic, predictable behavior" (Colten 2019, 417). He cites as an example the transition from hunting-gathering to agricultural lifestyles and sees transitions as having encompassed "numerous adaptations that thoroughly infused multiple aspects of society and demanded social, political, technological and economic transformations"(Colten 2019, 417). He therefore sees transitions as occurring at larger temporal and spatial scales than adaptations and as demanding transformations, whereas adaptations seek to perpetuate at least some aspects of the status quo. Given this tension between these two related processes, it is not surprising that he acknowledges that, in the case of coastal management in Louisiana since the arrival of the Europeans in the 1700s, "(m)ost adaptations were reactions to particular situations and not envisioned as a regional plan with a sustainable goal. Some have worked at cross purposes that created fundamental conflicts in current efforts to restore the coast" (Colten 2019, 430).

This observation is also applicable to the multiple adaptations that have taken place as individual components of the transition of Dandaragan's coastal plain from a locally self-sufficient seasonal hunting-gathering territory to a (post)modern and globally integrated multifunctional fishing, recreational and conservation zone. While Colten's observations can certainly be applied to the externally induced changes that have occurred in this coastal zone over time, conceptualizations of the frontier provide a more apt framework for the consideration of these changes in more local and spatial terms. The frontier is a colonial concept that distinguishes between a zone of disorder and a zone of order under the control of a state regime (Blomley 2003). While the concept of the frontier is strongly tied to the dispossession and misrepresentation of Indigenous land ownership and management (Blomley 2003), geographers have also demonstrated how it has been used in settler-colonial cities to displace Indigenous and other minority groups (Smith 1996; Launius and Boyce 2020), and in the spatial transformations of (allegedly) resource-rich rural areas through the concept of 'resource frontiers' (Barney 2009; Diniz 2019). Frontiers function as both a powerful ideological device that aligns with particular spatial transformations, often requiring legal changes and investment, and with the designation of frontier spaces that are contested zones of state and private initiatives and local livelihood practices.

In the case of the coastal zone of Dandaragan, change was muted for much of the period of European settlement and this area has therefore displayed frontier characteristics for an atypically long period of time. According to Alexandre M. A. Diniz (2019), these characteristics include being located "beyond or at least at the extreme limits of state control; thus, both these territories and the people inhabiting them lack effective formal or even sociocultural regulation". A relational approach to frontiers (Barney 2009) views these spaces as being actively peripheralized in a global economy, and the people and things within them as being caught up in a series of cumulative events that can usually transform, but at times maintain, their landscape and livelihood characteristics. In frontier areas, most inhabitants, and certainly the shack settlers of Wedge and Grey, experienced, and valued, the absence of formal regulation, and preferred their own informal management protocols. Simultaneously, however, frontier lands are often viewed from outside as spaces of opportunity where there is potential for economic gain and expanded political control (Barney 2009), both of which can be realized by means of land use adaptation or, perhaps more completely, by transition. Both this relational frontier perspective and that of adaptive transitions therefore encompass internal tensions; these tensions relate to the attitudes of both insiders and outsiders to the prospects and realities of preservation and change. Indeed, frontier insiders frequently contest external representations of their space and place as a 'frontier' (cf. Jacoby 2014; Griffin et al. 2019). Certainly, the Indigenous Yued people, who have used this zone for tens of thousands of years and perceive it as their home or, in their words, their 'country'/Boodja notwithstanding their historical dispossession, oppose the representation of this coastal plain as a frontier.

From another perspective, Holmes (2006) has referred to the changes currently taking place in rural Australia as a multifunctional transition in which intersections and confrontations occur between the values of production, consumption and protection. As a result of these interactions "increasing diversity, complexity and spatial heterogeneity in modes of rural occupance" (Holmes 2006, 144) are occurring, with seven generalized modes being identifiable in rural Australia. No fewer than four of these modes (Rural amenity, Pluriactive, Peri-metropolitan and Conservation and Indigenous) are applicable to the small area of the Dandaragan coastal plain with consumption, production and protection values either varying in importance or competing with each other in each instance. A variety of lenses therefore exist through which these local land use changes can be viewed.

Indigenous Occupation and Dispossession

McConnell et al. (1993, 7) observe that for over 40,000 years, the Dandaragan region contained a population which, "while it was essentially inward-looking, had learned to live in harmony with the land and the climate that governed it". This population, the Yued group of the Noongar Aboriginal people who inhabited the South West of Western Australia, adopted a localized and seasonal pattern of life (Hallam 2014), albeit while maintaining some trade and cultural links with adjoining and even distant Aboriginal bands. The Yued moved, regularly and systematically, between the coastal plain and the Dandaragan Plateau, some 50 kilometers inland. In spring, they moved from the plateau to the coastal plain where they would settle beside estuaries and swamps where fish and game were plentiful. In autumn, they would harvest supplies before moving inland for the winter. Evidence of Aboriginal occupance of the coastal zone was provided to the outside world by George Grey, the leader of the first European expedition to traverse this area in 1839. Grey (1841) noted "native villages" of "large well-built huts", areas where provisions had been harvested, and networks of tracks and wells. However, he crossed the coastal plain in April when, as he noted in his diary, "natives subsequently reported that the tract we have just traversed was at this season of the year totally devoid of water" (Grey 1841, np)—and also, therefore, of its Indigenous inhabitants.

Grey's party had "nearly perished from thirst and starvation" (McConnell et al. 1993, 13) on their journey and their negative reports on the environment and potential of the coastal plain echoed similar assessments of the area by Dutch shipwrecked sailors in 1658 (Henderson 1982) and French explorers in 1801 (Marchant 1982). However, the western third of Australia had been claimed for the British crown in 1827, when a military outpost was set up on the south coast at Albany and, in 1829, the Swan River Colony was established with Perth as its capital. The spread of settlement beyond Perth and the growth of the colony's non-Aboriginal population was slow, at least until convict transportation to Western Australia began in 1850 (Appleyard and Manford 1979). Nevertheless, by the 1840s, the first British shepherds were (illegally) grazing their flocks on land close to this area and, within a few years, the colonial authorities were formally granting pastoral leases over land which had hitherto been occupied by the Yued for a substantial part of the year (McConnell et al. 1993, 20).

Inevitably clashes, some of them fatal, occurred between the first European settlers in Western Australia and Noongar groups such as the Yued (Green 1984), but measles epidemics in the 1860s and 1880s and the spread of other diseases were a far more significant cause of deaths within the Indigenous population (Haebich 1988). Disease and dispossession therefore had a major impact on the ability of the Yued to maintain their traditional use of their country. Many of the Noongar families were relocated to missions. New Norcia was established in 1849 close to the eastern edge of what is now the Dandaragan Shire by Benedictine monks who sought to "Christianize and civilize"

Aborigines in this district and to establish a monastery surrounded by farming and pastoral land worked by Aboriginal families (Haebich 1988, 7). When the measles epidemic of the 1860s devastated the mission's Aboriginal population, the abbot refocused the mission's activities providing institutional care for Aboriginal children.

By the late nineteenth century, most Noongars were living on pastoral stations (sheep properties leased to British settlers on what remained as government land) and receiving subsistence allowances of food in exchange for domestic or station work. In the Road District (later Shire) of Dandaragan, Aboriginal people were "employed on all the larger properties, usually as stockmen or shearers" (McConnell et al. 1993, 105). However, they were "disempowered in so far as their traditional hunting grounds were concerned, assimilated to the point where many of them were by the beginning of the twentieth century products of mixed marriages yet never totally at ease with the ways imported from Europe, they continued to cling together as families, located at traditional meeting places" (McConnell et al. 1993, 106–7). Aboriginal historian Steve Kinnane (2003) views these nodes as bases of community resilience and resistance to settler colonial oppression, marginalization, and control. What might have been perceived as a mere local adaptation from an imperial (global) perspective was undoubtedly a transition for the local Yued population.

Settler Occupance of the Coastal Zone

Although pastoral leases were being granted and pastoral stations were being established about fifty kilometers inland on the Dandaragan Plateau in the mid nineteenth century, the coastal plain remained largely unused by the settler population for anything other than seasonal grazing of stock. A stock route, the Old North Road, which followed a line of swamps and connected Perth with the growing port of Geraldton and other settlements further north, was formally designated in the Government Gazette in May 1862 though it was only usable on a seasonal basis. With the opening of the land grant Midland Railway in 1894 (Figure 2) and a state government railway line still further to the east in 1915 (Higham 2007, 63–64) faster and more reliable routes between Perth and the north were opened up. Outside interest in this section of the coastal plain, which was, in any case, outside the railway's land grant zone, therefore declined in the early twentieth century. McConnell et al. (1993, 101–102) sum up the impact of the stock route's abandonment as follows:

"(T)he railway effectively reinforced the isolation of the Coastal Plain. Poison plants, deficient soils, a shortage of water and a rugged coastline ensured that the area remained largely unoccupied. ... In many ways, it underlined the frontier character of the few people who lived there; a character which occasionally bordered on the eccentric. ... It was an isolation, in fact, that enabled squatters, "roo" shooters, dog trappers and others who simply wanted to disappear, to live their lives in solitude."

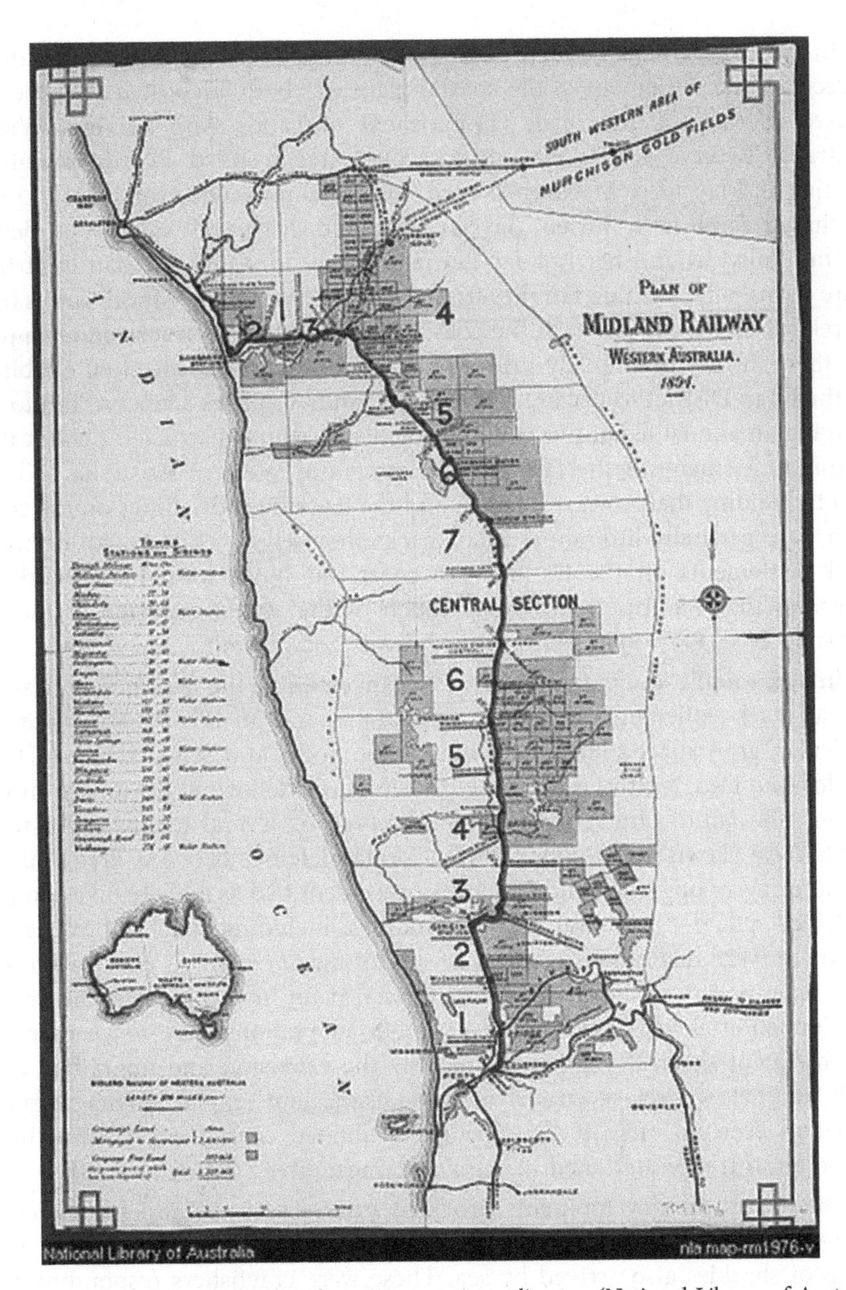

FIG. 2—Map of the Midland Railway Western Australia 1894 (National Library of Australia).

In short, the coastal plain offered its few inhabitants an escape from the outside world. This presaged many of the frontier characteristics exhibited, and even sought, by the impromptu holidaymakers who were to colonize the area in the coming decades.

In spite, or perhaps because of, this isolation and lack of services, the concept of recreational settlement of the coastal plain was both advocated and operationalized as early as the 1920s (Department of Lands And Surveys, Western Australia, Reserve 19206, File 2831/25; Dandaragan Road Board, Minutes of Meeting 18 May 1929; McConnell et al. 1993, 144–145). The Hon. Mary Lindsay purchased land near Jurien Bay in 1925 and 1927 and sought to develop a community around her holiday home and the store that she had built there. More informally, farming families from the Dandaragan and Moora shires began to trek (often for two days) to the coast along rudimentary tracks and camp and fish there in the summer. By 1929, this practice was sufficiently well established for the Road District to declare one area, at Sandy Cape, as a reserve and to start charging an annual fee of five shillings to anyone who constructed a shack there. At this time, many of the farmers and miners of Western Australia had been used to building their own residences and the few extant dwellings on the coastal plain were generally little more than shacks themselves. This growth of holiday shack settlements on the Dandaragan coast was but a local manifestation of a process that was occurring along much of the state's coastline at the time (Selwood et al. 1995; 1996; Selwood and Tonts 2004; 2006).

In the middle decades of the twentieth century, the number and size of coastal shack settlements in Western Australia, and on this stretch of coast in particular, grew apace (Suba and Grundy 1996; Jones and Selwood 2012). During World War Two, fears of a Japanese invasion grew following the attack on Pearl Harbor, the fall of Singapore and the bombing of several towns in Australia's North West (Lewis and Ingman 2013). Coastal defenses between Fremantle and Geraldton were upgraded and Jurien Bay was identified as a likely invasion point (Chappell n.d., 7–8). Several coastal buildings, including shacks at Wedge and elsewhere were demolished as part of a contingency plan to enact a scorched earth operation over the whole of the coastal plain in the event of an invasion (McConnell et al. 1993, 156). However, other adaptations were more supportive of subsequent shack development, notably the extension and upgrading of the track network to service coastal radar bunkers, gun emplacements, telephone lines and even an airstrip all of which facilitated coastal surveillance by the 20,000 or so troops stationed in the Dandaragan area (Jamieson 1978).

These wartime developments provided better access to and for the coastal shack settlement sites by land and, in the immediate post war period, a new group of shackies also arrived by sea. These were crayfishers responding to an explosion in the demand for Western Rock Lobster, initially from the United States and subsequently from Asia (Crombie 2001). At first, the fishers, from Fremantle and Geraldton, merely overnighted at safe anchorages along the coast, but the advantages of constructing shacks for their use during the fishing season was immediately apparent. From around 1950, crayfishers became an important component of the shack population at many of these settlements (Suba and Grundy 1996; Godden McKay Logan 2012). In two cases, at Jurien Bay in 1956 and Cervantes in 1963, these seasonal and hybrid holiday and commercial fishing settlements were formally gazetted as townsites where lots were surveyed for sale and the development of residences, fish processing facilities and even low order services was permitted. At this time, Grey was also surveyed as a possible townsite, but was not formally designated as such. Elsewhere, therefore, the shack settlements of both fishers and recreationists continued as illegal entities with no external controls over their modes of construction or operation (Figure 3).

More generally, this was the period of the post war long boom. Perth grew rapidly from a population of ca. 255,000 in 1940 to 680,000 in 1970 and Western Australia's population became more affluent, leisured and mobile. More locally, 'new land' for farming was opened up in Dandaragan, still on the plateau but closer to the coast (Carlin 1958). A new and more direct road from Perth to Geraldton, the Brand Highway (Route 1 in Figure 1), was opened through these farmlands in 1975.

Since it could still only be accessed by dirt tracks or along the Indian Ocean beaches, aspects of the frontier character of the coastal plain endured at least to the end of the twentieth century. However, the region was now considerably more accessible to considerably more people than hitherto and the shackie settlements therefore grew accordingly. Suba and Grundy (1996) noted that there were over 1,200 shacks along the central coast in 1994 and Wedge and Grey together comprised almost 25% of this total (Department of Planning and Urban Development 1994, Figure 11).

FIG. 3—The shack settlement of Grey (H. John Selwood 8/18/09).

Coastal Growth and Coastal Regulation: the Closing of the Frontier and the Evolution of New Modes of Occupance

Over the last half century or so, three developments, namely increased accessibility, rapid demographic and economic growth, and increased external environmental regulation, have served to diminish, if not totally remove, many of the frontier characteristics of the Dandaragan coastal plain. They have also served to bring consumption, production and protection values into conflict with each other. The section of the Brand Highway passing through Dandaragan Shire is only 30 kilometers from the coast; the former Perth-Geraldton road was twice as far inland. Other surfaced road connections were developed in the 1970s, notably links from Jurien Bay and Cervantes to the Brand Highway and from Perth to Lancelin, a fishing and holiday settlement located near Dandaragan's southern border and only 20 kilometers south of Wedge. Finally, a new coastal road, the Indian Ocean Drive (Route 60 in Figure 1), was opened in 2010 to facilitate tourism development along what is now termed, for marketing purposes, the 'Turquoise Coast' (Morrison et al. 2006) (Figure 4, which also shows the nature of the coastal plain's terrain).

These transport developments have facilitated rapid population growth at Jurien Bay and Cervantes which are now the largest settlements in the Shire with 2016 populations of 1,761 and 545 respectively. Since "only 50% of dwellings (in the shire) were occupied on census night" (Shire of Dandaragan 2020), which is in winter and thus outside both the crayfishing and the peak holiday seasons, the summer population of these centers would be significantly higher. The relocation of the shire council administrative offices from Dandaragan to Jurien Bay in

Fig. 4—The Indian Ocean drive shortly after its opening (Roy Jones 10/28/12).

2003 is perhaps the most telling indication that the shire's center of gravity has shifted from the primarily agricultural plateau to the hitherto overlooked coastal plain.

Inevitably, these population and accessibility pressures have had an impact on the natural environment of the coastal plain and here, as elsewhere in the state, this has led to increasing external intervention to protect aspects of the natural environment (Jones et al. 2007). State government intervention in this regard began on this stretch of coast in 1956 when Beekeepers Reserve was established. Three areas which, collectively, cover most of Dandaragan shire's coastal zone were gazetted as 'C' class reserves in 1972, 1973 and 1979. In 1994, these were amalgamated and given increased environmental protection as an 'A' class reserve called Nambung National Park. By the 1990s, over 80% of the coastline of Dandaragan Shire was classed as conservation reserve and 100% of its offshore area as a marine park. Nambung National Park, and particularly the impressive geological formations of the Pinnacles near Cervantes, has become a major tourist destination, with annual visitor numbers reaching 150,000 by 2011. From the perspective of the Department of Conservation and Land Management (1998, 1):

> Opportunities for sightseeing, bushwalking, nature appreciation, picnicking, coastal camping and four-wheel driving, all within a couple of hours drive from Perth, have meant large increases in visitor activity and commercial tourism in the area. Under their current tenure, many activities taking place in the nature reserves are not consistent with their primary purpose. These include camping, lighting of campfires, and off-road driving which causes a proliferation of tracks and degradation of dune vegetation. These problems are exacerbated by the presence of squatters' shacks at Wedge and Grey.

This exacerbation of problems resulting from the presence of the shack settlements had been noted by the State and local governments long before 1998 and their concerns extended beyond camping, fire lighting and off-road driving. The unregulated nature of the shack settlements and the lack of formal building standards have led to a variety of environmental and personal safety issues within an area where highly protected nature reserves now predominate. These include: the use of unsafe building materials, notably asbestos, in the construction of many of the shacks: the lack of adequate sewerage systems and resultant contamination of the local water table; and a range of building safety issues, notably the construction of shacks on unstable dunes too close to the coastal storm surge zone.

In 1968, State cabinet set up a committee to report on the unlawful use of Crown Land on the coast north of Perth. The resultant (Stokes) report in 1970 recommended the removal of all squatter settlements in the area and, in 1980, legislation was enacted "for the express purpose of providing the necessary means to remove squatters from public lands" (Suba and Grundy 1996). A statewide policy for the administration and, effectively, the removal, of coastal squatter shacks was adopted in 1988. However, the implementation of this policy was left to the small and poorly resourced coastal shires and shack removal therefore proceeded in an irregular manner. Gingin Shire, to the south of Dandaragan, had cleared all its shacks by the early 1980s and Coorow, to the north, by 1994. In Dandaragan, priority was initially given to the removal or upgrading of shacks in and around the formally designated townsites of Jurien Bay and Cervantes. However, after the Shire of Dandaragan had removed some of shacks under its control, tension developed between the Shire and the State government which, following lobbying by well-connected shack owners, largely withdrew its support for shack removal at Wedge and Grey.

The communities of Wedge and Grey were therefore left relatively undisturbed until most of the other shack communities had been removed. The effects of this were, first, that these two communities became a refuge for the more determined shackies who relocated, often with the building materials of their original shacks, from former settlements up and down the coast and, second, that they had time to devise strategies to resist or at least postpone their removal. Subsequently, the construction of Indian Ocean Drive effectively ended their isolation, and a state government report in 2011 recommended once more that the settlements at Wedge and Grey be dismantled. However the 'Cabinet Endorsed State Government Squatter Policy' of 1989 and 1999 remained in place and provided them with an interim arrangement whereby existing shack holders could be granted temporary tenancies in payment of an annual fee to cover the environmental rehabilitation and management costs that they placed on the shire and the state and these two settlements remain, at least for now.

Adaptation vs Transition: struggles for Survival on the Frontier

In the mid nineteenth century, the Yued Noongars' mode of occupance of Dandaragan's plateau and plain was brought to an end by, if not a globalizing then at least an imperial, incursion into their territory. Over a century later, the relatively localized land use and lifestyle practices of the shackies are likewise under threat from the external, and arguably global, forces of tourism, planning regulation (in particular the town planning schemes of local governments), and environmental conservation. While the circumstances and aims of the Indigenous and shackie groups differ considerably, they both continue to seek adaptations whereby at least certain aspects of their cultures and land use practices can avoid complete obliteration. There is also a small overlap between the two groups since some Yued people possess shacks at Wedge, though most now live elsewhere.

The shackie communities began to organize in order to preserve their lifestyles and the character of their settlements in the late 1960s when the Wedge

Island Progress Association was set up in 1968 and the Grey Conservation and Community Association in 1969 (Godden, McKay Logan 2012). By this time, the shack settlement communities had undergone a degree of generational change. This meant that not only did the shackie population at that time contain individuals who had experienced this recreational lifestyle from birth but also, and in line with the wider population, that a growing proportion of the shack owners came from educated, middle class, metropolitan backgrounds. Both these trends have continued over the last half century (Selwood and May 2001, 386). The former trend, together with the relocation of shackies from dismantled settlements elsewhere on the coast, may well have contributed to a growing determination on their part to preserve as much of their unrestricted recreational lifestyle as possible. In our discussions with the shackies and in the submissions made to government inquiries, frequent references were made to the multigenerational nature of shackie life, to its formative impact on children in terms of their environmental and social awareness and to the shackies' affinity to and concern for the land that they occupied. There are differences between the communities of Wedge and Grey, including how the more affluent and better-connected families at Grey have shown a greater willingness to compromise with regard to government regulation of their settlement. However, both shack associations have, for decades now, contained individuals with the expertise to negotiate skillfully with the external authorities over their future and to promote their cause within the wider community through the media. Among our interlocutors at Wedge and Grey were a retired judge and several other professionals and the community associations' flyers and public submissions display high levels of political, social and media sophistication.

In partial response to the 1989 Squatter Policy, and following discussions with the shackie population, a visiting academic, John Selwood, prepared a discussion paper for the Department of Planning and Urban Development (Selwood 1991). This paper recommended that some of the shack settlements be allowed to remain in a modified form and on a more secure leasehold basis. It also raised, for the first time, the issue of the heritage value of the shack settlements. Although this document was promptly designated "Draft only. Recommendations not accepted" and "Closed Reserve only. Not endorsed for public comment" by the Central Coast Planning Study Steering Committee, the issue of heritage value was enthusiastically taken up by the shackie organizations themselves. They noted that some shack settlements elsewhere in Australia were being accorded heritage protection (New South Wales National Parks and Wildlife Service 1994) and, through a voluntary regional group, Midwest Heritage Incorporated, funding was obtained from the National Estate Grants Program of the Australian Heritage Commission for a survey of the shacks on the Central Coast of Western Australia. This survey (Suba and Grundy 1996) focused on the shacks in the shires immediately to the north of Dandaragan since these were in the process of removal at that time.

Suba and Grundy's report therefore sought to record, through photographs, oral histories and document analysis, what appeared to be a disappearing way of life. Its Executive Summary contended that "the shacks have aesthetic value as a group which forms a unique cultural landscape, historical significance for their association with their opening up of the coastline and social value for their representation of a disappearing lifestyle". Indeed, this elegiac aspect of the report found brief expression in a beachside interpretive display board depicting shackie history which was set up at a former squatter settlement (Selwood and May 2001, 388). However, the shackies at Wedge and Grey refused to accept their imminent demise and continued to campaign and to lobby government for their right to remain. Discussions over the fate of these two settlements have ebbed and flowed over recent decades. The Wedge and Grey Draft Master Plan (Department of Conservation and Land Management 1998) envisaged a progression toward more mainstream recreational uses of these locations which would nevertheless retain some of the esthetic and unregulated aspects of the shackie lifestyle. By contrast, the WA Legislative Council Standing Committee on Environment and Public Affairs (Western Australian Legislative Council 2011), in cognizance of the opening of the Indian Ocean Drive in the previous year, recommended the total removal of both settlements.

To date, state parliament has not accepted this terminating recommendation. Rather, it has noted, but not ruled on, the findings of another heritage report (Godden Mackay Logan 2012). This report was commissioned by the National Trust of Australia (WA) and takes a more proactive stance than that of Suba and Grundy by recommending the inclusion of the Wedge and Grey shack settlements on the Western Australian Register of Heritage Places and requesting "the State Government Minister for the Environment (to) develop a holistic management plan which conserves and interprets the cultural heritage values of these precincts". To this end, the Wedge and Grey community associations used the shack settlements' newfound accessibility to organize an Open Day to publicize the report's recommendations and to leverage wider community support (Figure 5). Subsequently, they have produced a draft "Wedge and Grey Planning Framework" (Lutton and Chessels 2018) and are currently working with the state government on a regulatory framework for the management of these settlements as diversified recreational sites with a mix of camping, caravanning and shack components (Department of Biodiversity, Conservation and Attractions 2019).

The spatial regulation of this frontier zone continues to be a work in progress, a process which has been further complicated, over the last half century, by a series of changes with regard to Aboriginal rights and access to land. While these developments largely took place at the federal and state levels, they have had local implications on the Dandaragan coast. In 1972, Western Australia passed the Aboriginal Heritage Act. This gave the state the authority to designate and the power to protect significant Aboriginal sites. Despite

FIG. 5—Heritage messaging at the Grey Open Day (Roy Jones 10/28/12).

a weakening of these protections over time (Herriman 2013), the Aboriginal significance of sites needs to be taken into consideration in the state's planning and development processes.

In a related, but parallel, development in 1992, a case brought by Eddie Mabo against the Queensland government had the effect of overturning the hitherto federally acknowledged doctrine of terra nullius, namely that the land was unoccupied when the British claimed possession of Australia in 1788. As a result of this decision, the federal government passed the Native Title Act of 1993. This Act established the Native Title Tribunal, a body which has the power to assess claims by what are termed the Traditional Owners for the restitution of any land not already in private freehold ownership. These claims are prepared and submitted by a series of regional Aboriginal Land Councils who represent the claimants. The South West Land and Sea Council (SWALSC) is responsible for progressing claims by Yued people and five other Noongar groups. In 2015, it succeeded in bringing about the South West Native Title Settlement. This agreement between the Western Australian government and SWALSC is the largest Native Title settlement in Australia to date. It provides for land allocations and financial compensation to the Noongar nation in return for the relinquishment of further Native Title claims, and Wedge and Grey meet the criteria to be part of this allocation. While appeals over the legality of the Settlement have delayed its implementation, the broader growth of Aboriginal-state co-management of nature reserves since 2012, where Aboriginal custodians and state government departments are management partners, is already shifting relationships at Wedge and Grey. Given the probability that joint management will be implemented in the Nambung National Park, and therefore over Wedge and Grey, the state Department of Biodiversity, Conservation and Attractions is developing strong relationships with Yued leaders.

The Wedge and Grey sites are on Crown (i.e. public) Land and are thus liable to Native Title claim. They are "areas of significance to the Yued and form part of a traditional song-line that extends from Dongara to Yanchep. Wedge and Grey were important locations for the Yued for thousands of years as a meeting place, for camping and ceremony, and as a food and water resource" (Department of Biodiversity, Conservation and Attractions 2019, 8–9). Wedge contains one of the most significant Aboriginal midden sites in the South West of Western Australia, and there are several registered Aboriginal sites in the vicinities of both Wedge and, to a lesser extent, Grey. Although these sites are already under protection, they have experienced some damage and disruption as a result of the development of four-wheel drive tracks and other shackie activities. This is a matter of concern for at least some of the Yued population. Nevertheless, the reality of historic dispossession for the Yued and the threat of future dispossession for the Wedge and Grey shackies provide both groups with some basis for cooperation if they wish to retain at least partial access and use rights to this coastal land. Equally, the territorial overlapping of Yued and shackie heritage, and the increasing opportunities for Yued to influence land management decisions at Wedge and Grey, present potential flash points for community conflict. At Wedge where the more significant Aboriginal heritage sites are located (Figure 6), the Yued have recently registered complaints with the relevant minister over the behavior of some shackies.

Conclusion: Adaptations and Transitions on A Closed Frontier

In his consideration of coastal transitions and adaptations, Colten (2019) draws on the work of Neil Adger (Adger 2000; Adger et al. 2005, 2009) on resilience and, specifically, on "the ability of communities to withstand external shocks to their social infrastructure" (Adger 2000, 361). Adger's work is largely focused on the external shock of climate change but his argument that "limits to adaptation are mutable, subjective and socially constructed" (Adger et al. 2009, 338) is also applicable to the ability of communities to withstand other forms of external shock. For the Yued at the time of colonization, the externally imposed limits to their adaptation were extremely narrow and, in the medium term at least, they experienced a brutal and relatively immediate transition from a local and autonomous to a subaltern and colonial way of life. The shack communities of

FIG. 6—Delimitation of Indigenous land at Wedge (Tod Jones 4/22/19).

Wedge and Grey also experienced and have strongly valued their local and autonomous way of life, albeit for decades, rather than for tens of millennia. While the surrounding shack settlements have recently experienced a transition to oblivion, to date the Wedge and Grey shackies appear to have retained at least some potential to adapt to their changing external circumstances.

At one level, therefore, the Yued and the shackies can be said to have experienced comparable external and even global threats to their existence. Indeed Jacoby (2014, 202) discerns a frontier sequence whereby "(t)he ensuing replacement of one population by another produces a series of inversions in which the settlers become the new natives while the indigenous peoples become outsiders in their former homeland". As the frontier on the Dandaragan coastal plain first expanded and then closed, both the Yued and the shackies have transitioned from being relatively autonomous spatial insiders to being social outsiders in an increasingly regulated and globally integrated coastal zone in which conflicts between consumption, production and protection values are growing. However, these commonalities of experience have not, as yet, led them to work together to any significant extent on achieving "an overarching regional plan with a sustainable goal" (Colten 2019, 430). Indeed, as both groups have experienced "adaptations (that) were reactions to particular situations" they have invoked national and even international parallels and policies in order to enter into debates on Indigenous and settler heritage and therefore over which of these heritages should prevail over the other, or indeed over the area's natural heritage. As such, they have "worked at cross purposes that created fundamental conflicts in current efforts to restore the coast" (Colten 2019, 430).

It is, perhaps, ironic that the terms of the South West Native Title Settlement have the potential for the Yued to claim the Wedge and Grey reserves in their entirety and, apparently thereby, to exercise considerable control over the management of both shack settlements and to impose a 'conservation and Indigenous' mode of occupance over much of the coastal plain. This is part of a significant change in land use driven by the introduction of joint management between the Department of Biodiversity, Conservation and Attractions and the Aboriginal Traditional Owners following legislative amendments in 2012. On the surface, this would seem to take these local processes of land use transition full circle, thereby ending the frontier status and possibly even the existence of the shack settlements. However, in a post frontier situation, the state government retains considerable powers over the area's environmental conservation, planning regimes, and economic development. A planning submission on behalf of the Yued group in 2014 and subsequent documents indicate that they would be prepared to leave responsibility for managing parts of the shack settlements to the State government while seeking joint management over land that included their significant heritage sites. While this submission appears to leave room for compromise, it also assumes the removal and relocation of a number of shacks on land significant to the Yued, some restrictions on shackie land use practices and a degree of Yued management of visitors, all of which are currently opposed by the Wedge Island Progress Association. Whether these competing groups can now work together to ensure rural sustainability and vitality on the Dandaragan coastal plain in a new equilibrium, as envisioned by Diniz (2019), or whether they fail to adapt as a result of fundamental conflicts of the type feared by Colten (2019) remains to be seen. These disputes are local manifestations of global trends in population distribution, peri-urban coastal development, environmental protection and Indigenous recognition that have occurred in an area which, for much of the last two centuries, received little or no economic or infrastructural investment or attention. As such, they illustrate some of the challenges in achieving rural sustainability and vitality at a local scale in a globalizing world.

Acknowledgments

In memoriam, Professor H. John Selwood, University of Winnipeg (1936-2020) in recognition of his pioneering research on informal settlements in Western Australia and elsewhere.

References

Adger, W. N. 2000. Social and Ecological Resilience. Are They Related? *Progress in Human Geography* 24 (3):347–364. 10.1191/030913200701540465

Adger, W. N., S. Desai, M. Goulden, M. Hulme, I. Lorenzoni, D. R. Nelson, O. Naess, J. Wolf, and A. Wreford. 2009. Are There Social Limits to Adaptation to Climate Change? *Climatic Change* 93 (3):335–354. 10.1007/s10584-008-9520-z

Adger, W. N., T. P. Hughes, C. Folke, S. R. Carpenter, and J. Rockstrom. 2005. Social-ecological Resilience to Coastal Disasters. *Science* 309 (5737):1036–1039. 10.1126/science.1112122
Appleyard, R. T., and T. Manford. 1979. *The Beginning: European Discovery and Early Settlement of the Swan River, Western Australia*. Nedlands WA: University of Western Australia Press.
Barney, K. 2009. Laos and the Making of a 'Relational' Resource Frontier. *The Geographical Journal* 175 (2):146–159. 10.1111/j.1475-4959.2009.00323.x.
Blomley, N. 2003. Law, Property, and the Geography of Violence: The Frontier, the Survey, and the Grid. *Annals of the Association of American Geographers* 93 (1):121–141. 10.1111/1467-8306.93109.
Carlin, B. F. 1958. Light Lands Development in the West Midlands. *Journal of the Department of Agriculture Western Australia* 7:59–60.
Chappell, J. n.d. *Again the Hounds of War are Loosed. How Western Australia Responded*. Perth: Battye Library, State Library of Western Australia.
Colten, C. E. 2019. Adaptive Transitions: The Long-term Perspective on Humans in Changing Coastal Settings. *Geographical Review* 109 (3):416–435. 10.1111/gere.12345
Crombie, G. I. 2001. The Influence of Technological Change upon the Social Organisation and Geographical Location of the Western Australian West Coast Fishing Industry since 1950. PhD thesis, Curtin University of Technology.
Crowley, F. K., ed. 1970. *Australia's Western Third: A History of Western Australia from the First Settlements to Modern Times*. Melbourne: Heinemann.
Darwin, C. 1859. *On the Origin of the Species by Means of Natural Selection or the Preservation of Favoured Races in the Struggle for Life*. London: John Murray.
Department of Biodiversity, Conservation and Attractions. 2019. *Wedge and Grey Draft Policy Manual and Maintenance Guidelines*. Perth: Department of Biodiversity, Conservation and Attractions.
Department of Conservation and Land Management. 1998. *Nambung National Park Management Plan 1998–2008. Management Plan No. 37*. Perth: Department of Conservation and Land Management for the National Parks and Nature Conservation Authority.
Department of Planning and Urban Development. 1994. *Central Coast Regional Profile*. Perth: Department of Planning and Urban Development.
Diniz, A. M. A. 2019. Conceptualizing the Frontier. In *Twentieth Century Land Settlement Schemes*, edited by R. Jones and A. M. A. Diniz, 11–27. London and New York: Routledge.
Drew, P. 1994. *The Coast Dwellers: A Radical Reappraisal of Australian Identity*. Ringwood, Victoria: Penguin.
Green, N. 1984. *Broken Spears: Aboriginals and Europeans in the Southwest of Australia*. Perth: Focus Education Services.
Grey, G. 1841. *Journals of Two Expeditions of Discovery in North West and Western Australia*. London.
Griffin, C., R. Jones, and I. Robertson, eds. 2019. *Moral Ecologies. Histories of Conservation, Dispossession and Resistance*. London: Palgrave Macmillan.
Haebich, A. 1988. *For Their Own Good: Aborigines and Government in the Southwest of Western Australia 1900-1940*. Nedlands WA: University of Western Australia Press.
Hallam, S. J. 2014. *Fire and Hearth: A Study of Aboriginal Usage and European Usurpation in South-Western Australia. Revised Edition*. Nedlands, WA: University of Western Australia Publishing.
Hardy, D., and C. Ward. 1984. *Arcadia for All: The Legacy of a Makeshift Landscape*. London and New York: Mansell.
Henderson, J. A. 1982. *Marooned: The Wreck of the Vergulde Draeck and the Abandonment and Escape from the Southland of Abraham Leeman in 1658*. Perth: St George Books.
Herriman, N. 2013. Western Australia's Aboriginal Heritage Regime: Critiques of Culture, Ethnography, Procedure and Political Economy. *Australian Aboriginal Studies* 2013(1):85–100.
Higham, G. 2007. *Marble Bar to Mandurah: A History of Passenger Rail Services in Western Australia*. Bassendean WA: Rail Heritage WA.
Holmes, J. 2006. Impulses Towards a Multifunctional Transition in Rural Australia. Gaps in the Research Agenda. *Journal of Rural Studies* 22 (2):142–160. 10.1016/j.jrurstud.2005.08.006

Jacoby, K. 2014. *Crimes against Nature: Squatters, Poachers, Thieves and the Hidden History of American Conservation with a New Afterword.* Berkeley, Los Angeles and London: University of California Press.

Jamieson, W. D. 1978. *A History of the Army in Western Australia in World War II.* Perth: Australian Army.

Jones, R., C. Ingram, and A. Kingham. 2007. Waltzing the Heritage Icons: 'Swagmen', 'Squatters' and 'Troopers' at North West Cape and Ningaloo Reef. In *Geographies of Australian Heritages: Loving a Sunburnt Country?*, edited by R. Jones and B. J. Shaw, 79–94. Aldershot: Ashgate.

Jones, R., and H. J. Selwood. 2012. From Shackies to Silver Nomads: Coastal Recreation and Coastal Heritage in Western Australia. In *Heritage from Below*, edited by I. Robertson, 125–146. Farnham: Ashgate.

Kinnane, S. 2003. *Shadow Lines.* Fremantle: Fremantle Press.

Launius, S., and G. A. Boyce. 2020. More than Metaphor: Settler Colonialism, Frontier Logic, and the Continuities of Racialized Dispossession in a Southwest U.S. City. *Annals of the American Association of Geographers*:1–18. 10.1080/24694452.2020.1750940.

Lewis, T., and P. Ingman. 2013. *Carrier Attack Darwin 1942: The Complete Guide to Australia's Own Pearl Harbor.* Kent Town, SA: Avonmore Books.

Logan, G. M. 2012. *Wedge and Grey Shack Settlements Cultural Heritage Assessment: Report Prepared in Collaboration with Context Pty Ltd for National Trust of Australia (WA).* Sydney and Canberra: Godden Mackay Logan Pty Ltd.

Lutton, L., and M. Chessels. 2018. *Wedge and Grey Planning Framework January 2018 Draft.* Perth: Wedge Island Protection Association and Grey Conservation and Community Association.

Marchant, L. R. 1982. *France Australe: A Study of French Explorations and Attempts to Found A Penal Colony and Strategic Base in South-Western Australia 1503-1826.* Perth: Artlook.

McArthur, W. M., and E. Bettenay. 1960. *The Development and Distribution of Soils of the Swan Coastal Plain Western Australia: CSIRO Soil Publication 16.* Melbourne: CSIRO.

McConnell, M., J. McGuire, and G. Moore. 1993. *Plateau, Plain and Coast: A History of Dandaragan.* Dandaragan: Shire of Dandaragan.

Morrison, S., A. Storrie, and P. Morrison. 2006. *The Turquoise Coast.* Kensington, WA: Department of Conservation and Land Management.

New South Wales National Parks and Wildlife Service. 1994. *Royal National Park Cabins Conservation Plan. Draft. September 1994.* Sydney: New South Wales National Parks and Wildlife Service.

P.I.R.G. 1977. *A Coastal Retreat.* Clayton, Victoria: Victorian Public Interest Research Group.

Selwood, H. J. 1991. *Squatters on the Central Coastline of Western Australia: A Discussion Paper with Recommendations.* Perth: Department of Planning and Urban Development.

———. 2006. The Evolution, Characteristics and Spatial Organization of Cottages and Cottagers in Manitoba, Canada. In *Multiple Dwelling and Tourism: Negotiating Place, Home and Identity*, edited by N. McIntyre, D. R. Williams, and K. E. McHugh, 207–218. Wallingford, Oxfordshire: CAB International.

Selwood, H. J., G. Curry, and R. Jones. 1996. From the Turnaround to the Backlash: Tourism and Rural Change in the Shire of Denmark, Western Australia. *Urban Policy and Research* 14 (3):215–225. 10.1080/08111149608551597

Selwood, H. J., G. Curry, and G. Koczberski. 1995. Structure and Change in a Local Holiday Resort: Peaceful Bay on the Southern Coast of Western Australia. *Urban Policy and Research* 13 (3):149–157. 10.1080/08111149508551649

Selwood, H. J., and A. May. 2001. Resolving Contested Notions of Tourism Sustainability on Western Australia's 'Turquoise Coast'. *Current Issues in Tourism* 4 (2–4):381–391. 10.1080/13683500108667894

Selwood, H. J., and M. Tonts. 2004. Recreational Second Homes in the South West of Western Australia. In *Tourism, Mobility and Second Homes: Between Elite Landscape and Common Ground*, edited by C. M. Hall and D. Muller, 149–161. Clevedon: Channel View.

———. 2006. Seeking Serenity: Homes Away from Home in Western Australia. In *Multiple Dwelling and Tourism: Negotiating Place, Home and Identity*, edited by N. McIntyre, D. Williams, and K. McHugh, 161–179. Wallingford: CAB International.

Shire of Dandaragan. 2020. Strategic Plan 2016–2026 https://www.dandaragan.wa.gov.au/documents/1337/strategic-community-plan-2016-2026

Smith, N. 1996. *The New Urban Frontier: Gentrification and the Revanchist City*. New York: Routledge.

Stannage, C. T., ed. 1981. *A New History of Western Australia*. Nedlands, WA: University of Western Australia Press.

Suba, T., and G. Grundy. 1996. *The Survey of Squatter Shacks on the Central Coast of Western Australia*. Canberra: Australian Government Printing Service.

Wallerstein, I. 1976. From Feudalism to Capitalism: Transition or Transitions? *Social Forces* 55 (2):273–283. 10.2307/2576224

Western Australian Legislative Council. 2011. *Standing Committee on Environment and Public Affairs Report 21 Shack Sites in WA*. Perth: Western Australian Legislative Council.

EXPLORING THE SUSTAINABILITY OF WILDERNESS NARRATIVES IN EUROPE. REFLECTIONS FROM VAL GRANDE NATIONAL PARK (ITALY)

GIACOMO ZANOLIN and VALERIÀ PAÜL

ABSTRACT. *Wilderness* is an important concept in the discourses and policies concerning contemporary European protected areas, inherently challenging in terms of sustainability. Since its designation in 1992, the Val Grande National Park, located in northwest Italy, has been portrayed and promoted as a wilderness area, thereby enhancing tourism, whilst disregarding the historical signs of human activity. In this paper we explore the wilderness concept, focusing on the narratives developed in the area, that changed from a strict conservationist approach to a more utilitarian one, influencing the National Park's policy-making. The research is based mainly on the content analysis of several literary texts. We conclude that *wilderness*needs to be reconceptualized so that contemporary European protection policies might become more effective, and we may use our knowledge of nature to promote sustainable development.

The concept of wilderness, as exported from the pioneering United States National Park system since the early twentieth century (Nash 1964 [ed. 2014], Nash 1970), is difficult to convey in Europe because the areas protected as "wild nature" disregard the long-term interventions made by humans over the centuries, thus rendering their management very complex (Dudley 2011; Woods 2011). Hence, the word "wilderness" in European policy making is a challenge for sustainability, since this latter concept suggests a balance between nature conservation and economic activities.

In the specific case of the Val Grande National Park (Piedmont, Italy) (Figure 1), the word wilderness, in English, has been used at least since the 1970s. Since its designation as a national park in 1992, it has been characterized as the "biggest wilderness area in the Italian Alps." This label, based on a peculiar social construction of nature (Castree 2001), has been extensively used as a local development strategy, resulting in the enhancement of a tourism dimension. A case in point is the achievement, in 2013, of the European Charter for Sustainable Tourism in Protected Areas (ECST), confirming the correlation between wilderness and the sustainability arena in Val Grande.

The aim of this paper is to consider how the notion of wilderness has developed in a European context, examining its ties with the notion of sustain-

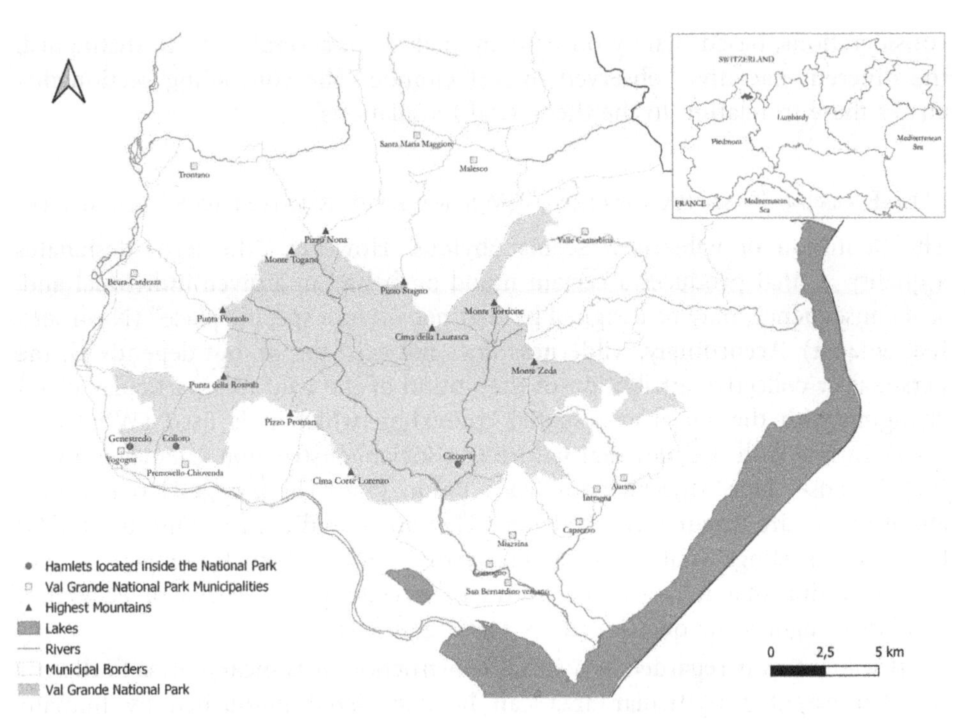

FIG. 1—Location map. Source: Giacomo Zanolin, 2020.

ability. Wilderness apparently signifies a lack of human intervention for the purposes of nature conservation, but in Val Grande the promotion of tourism strongly based on this concept allows us to question to what extent it can be reframed under the lens of sustainability. Indeed, Val Grande has been promoted as a unique pristine natural area within the Alpine region, comprising various forms of economic activities that seem to be in contradiction with the wilderness notion itself but that are, at least since Peter Nijkamp (1990), inherently part of the sustainability paradigm. We are not the first to study the Val Grande National Park under a wilderness glass. Franz Höchtl (2005, 2006, 2007) and Claudia Cassatella (2016) focused on land-use/land-cover changes since the nineteenth century and visitor perception. However, our aim here is not to determine to what extent the region can be considered wild from an "objective" perspective, or whether it is seen as wild by tourists. Instead, we will deal with the narratives that have emerged about the area, in order to elucidate why it is seen as a wilderness, by whom, with what interests, and with what consequences in relation to sustainable development strategies. In this sense, we assume wilderness to be a social imaginary of space as defined by Bernard Debarbieux (2019) and in Val Grande we examine a specific application of this socio-spatial construction and its inherent policy implications.

The paper begins by conceptualizing wilderness as a socio-spatial construction, and goes on to present the case-study area and explain the methodological considerations, based mainly on content analysis of several texts, to distinguish the different narratives observed in Val Grande. The concluding section discusses these in relation to the theoretical foundations.

The Evolving Conceptualization of Wilderness in Relation to Sustainability

The definition of wilderness seems obvious. However, "the term designates a quality ... that produces a certain mood or feeling in a given individual and, as a consequence, may be assigned by that person to a specific place" (Nash 1967 [ed. 2014], 1). Accordingly, wilderness does not exist per se, but depends on the personal or collective attribution of the notion of the wild to a particular place, changing with the social and spatial context in which it is used (Whatmore 2002), in line with the idea that nature is a social construction (Demeritt 2002). In this sense, Jane Moeckli and Bruce Braun (2001) have argued that claims about nature are discursively mediated. Therefore, wild nature could be studied by deconstructing what it is known about nature, "denaturalizing" it and demonstrating that it is a social product "serving specific social or ecological ends that ought to be questioned" (Castree 2001, 13).

If wilderness is regarded as a social construction, it is meaningless to discuss to what extent a particular area can be considered untouched by humans (Dudley 2011). It is evident that "the original, natural landscape ... [i]n its entirety ... no longer exists in many parts of the world" (Sauer 1925, 37). Haraway's (1988) criticisms state that the assumption that human agency and nature are binary opposites is a Western notion. She argues that, "Nature is only the raw material of culture, appropriated, preserved, enslaved, exalted, or otherwise made flexible for disposal by culture in the logic of capitalist colonialism" (Haraway 1988, 592), a point that has been also contended by Neil Smith (2007).

The origin of the wilderness notion is ancient. It derives, possibly, from mythical and religious visions of fearful, uninhabited, and uncivilized places, like deserts or mountains, where gods, deities, and evil could live together; this is what Yi-Fu Tuan suggests in his texts from the 1970s (Nogué 2018). Wilderness gained a general recognition in England in the late fourteenth century, with the English translation of the Bible (Meli 2007). In this period, the concept was defined to describe, beyond the blessed farmlands, those areas that were uncontrolled and inhabited by wolves and other wild animals with a strong negative, archetypical, cultural significance that stimulated the "fear of the wilderness" (Short 2006; Woods 2011). In the centuries that followed, wilderness was used in various texts to describe treeless wastelands. The term migrated from Europe to America in the seventeenth century, where it was used to refer to the vast spaces that were uncommon in European landscapes (Nash 1967 [ed. 2014]).

However, during the periods of Enlightenment and Romanticism (in the late seventeenth and early eighteenth centuries) feelings about nature underwent a deep transformation. Romanticism fabricated a wild nature and promulgated a taste for it from a biocentric perspective (Descola 2015). In particular, a fresh interest in the mountains was fostered (Martínez de Pisón and Álvaro 2002; De Rossi 2014) and, in a broader sense, in uncivilized spaces described as wild (Brevini 2013). In the Alps especially, a new perspective developed, whereby the mountains became the esthetic and emotional paradigm. Then, the myth of the *bon sauvage* proposed by Rousseau permeated European culture, focusing attention on the "wild subjectivity" of people who lived in the Alps, as opposed to the urbanized and civilized people living in the lowlands (Salsa 2009).

The new notion of wilderness found a specific and fundamental construction in the United States where the pioneering, frontier life in the expanding West, gradually led to an idea of the wilderness as a setting where personal aspirations for introspection and transcendence could be projected (Tuan 1996; Depraz 2008). Wilderness became the environment where it was still possible to experience feelings of loneliness, harmony, and peace, which had been deemed lost in the chaos of urbanization and industrialization. It also became something akin to a state of mind, a way of living with nature and an opportunity to reconcile human beings with themselves (Tuan 1996; Nogué 2018).

Within this perspective, wilderness was converted into a concept to promote the preservation and enjoyment of natural areas as opposed to the urban and industrial development of the United States in the nineteenth century. It became the key concept for the preservationist movement, led by Muir among others, that promoted protected areas, beginning with the designation, in 1864, of the first nature reserve in the Yosemite Valley and, in 1872, of the first national park in Yellowstone (Nash 1970; Depraz 2008; Frost and Hall 2009; Debarbieux 2019). Hence, in the United States, the modern idea of wilderness was forged for two main purposes: to build social consensus around the designation of protected areas to safeguard a set of values, and to highlight the contrast with Europe of New World uniqueness, underlining the existence of something purely "American" (Nash 1967 [ed. 2014]). Significantly, research by authors such as Jacoby (2001) has shown that, when it was designated, Yellowstone was not an untouched nature area at all, but it has become important for the generated spatial imaginary in Debarbieux's (2019) terms.

Accordingly, wilderness was associated with nationalism at the end of the nineteenth century, as was evident in the United States (Depraz 2008). Yi-Fu Tuan (1996) has explained how wilderness became a political concept, indicating that its creation was assumed to be "natural" in the Western perspective, though nonexistent in Chinese culture. The translation to other countries of the notion of wilderness, through preservationist policies, was highly problematic. For instance, in Europe, in the early twentieth century, mountains and forests were rapidly identified with clear nationalist connotations (Depraz 2008).

The notion of wilderness has been reinvented over the years (Brevini 2013) owing to the transformation of the protectionist movement throughout the twentieth century and the rise of new paradigms. John Muir's seminal ideas somehow progressed through authors such as Leopold, the founder of contemporary environmentalism. In parallel, Pinchot's conservationist notion emerged. Conservation advocates that resources can be efficiently exploited by humans (Callicott 1998). Hence, the distinction between preservationism and conservationism has emerged as a basic dualism in contemporary environmental thinking, planning, and management (Oelschlaeger 1991; Depraz 2008). According to Juan F. Ojeda (2006), sustainability typically links with conservationism in the sense that is spatially selective: only some areas are ambiguously designated as protected in order to develop specific sustainability targets, while the others are not included from the viewpoint of sustainability.

These discussions are related to the aforementioned Western schism between humans and nature (Haraway 1988; Smith 2007; Descola 2015). Samuel Depraz (2008) and Matteo Andreozzi (2017) argue that there are three main environmental ethics. First, preservationism is related to biocentric ethics, which extends the role of moral agent to all living creatures. In general terms, this is in line with deep ecology (Naess 1973; Capra 1995) and with the new environmental paradigm (Dunlap and Van Liere 1978; Van Den Born 2001), both originating in the 1970s, and advocating for strict wilderness preservation on the basis that mankind is only one of many species that inhabit the Earth. Second, conservationism is ethically ecocentric in the sense that nature can be partially used in conjunction with environmental protection in particular areas. Finally, anthropocentric ethics believe humans are the fulcrum, so nature must be at their disposal.

The concept of sustainability, as developed since the late 1980s, is intrinsically anthropocentric given that it equates the environment to the human agency, and attempts to find a balance, using the well-known three-dimensional model, in which sustainability is described as the combination of economy, society, and the environment (Nijkamp 1990; Blewitt 2008; Borowy 2018). However, there is some confusion between sustainable development and growth (Ojeda 2006; Naredo 2007), and what can potentially be sustainable is the former: growth is mostly linked to the economic dimension of sustainability (that is, GDP growth), while development is supposed to reconcile the economy with environmental, social, and cultural dimensions since sustainable development was seminally defined by the World Commission on Environment and Development (1987) in the so-called "Brundtland Report."

A case in point is the so-called Great Wilderness Debate instigated by two miscellaneous volumes edited by J. Baird Callicott and Nelson (1998) and Michael P. Nelson and Callicott (2008). The editors call for a new understanding of wilderness in two directions: de-anthropocentrizing the classical idea of wilderness by creating biodiversity reserves safeguarded from human intervention, and reconceptualizing wilderness to remove any human perception, implying that it exists per se independently from the human viewpoint. The former direction clearly links with Callicott's (1998) understanding of sustainable development, which proposes to limit "economic activity ... by ecological exigencies; ... not seriously compromis[ing] ecological integrity; and, ideally, ... positively enhanc[ing] ecosystem health."

Over the last few decades, sustainability has been frequently linked to tourism, morphing into the definition of sustainable tourism. According to Hughes and others (2015), this widely used term has attracted considerable attention from researchers and policy makers, in spite of a wide divergence between theory and practice that generates many paradoxes. As regards protected areas, sustainable tourism has become a synonym of tourism in natural areas or ecotourism (Hunter 2002; Frost and Hall 2009), creating obvious paradoxes, especially when referring to (alleged) wilderness areas. The wilderness notion has attracted considerable tourist attention since the seminal national park designations of the nineteenth century (Frost and Hall 2009). According to John Urry (2002), the so-called Romantic tourist gaze is based on the sensations of loneliness and privacy, especially attainable in a wilderness context. The opposing tourist gaze would be the collective, and requires a large number of people. Apparently, this cannot be developed in wilderness contexts. However, wilderness areas, above all those that have been designated and protected, have also been modified and adapted for sustainable tourism activities (Hunter 2002) and tourist enjoyment (Lovell and Bull 2018). In the end, if these efforts succeed, then "the more tourists are attracted to them, the less authentic they become" (Lovell and Bull 2018, 130).

Based on these theoretical insights, this paper assesses how these variations occurred in Val Grande National Park. The paradoxes and tensions seem evident in the case-study area because of the continued reiteration of two concepts: wilderness and sustainable tourism. According to the literature reviewed, wilderness is a biocentric concept and sustainability an anthropocentric one, thus implying a contradiction.

Case Study Area

Designated in 1992, the Val Grande National Park is one of the most recent in Italy (Figure 1). It consists in area of 14,598 hectares located in northwest Piedmont, near the border with Switzerland. Val Grande is the name of the principal valley, but nowadays the designation has extended to the whole national park area. The other main valley in the park is named Pogallo. The Valgrande River becomes the San Bernardino River at the confluence with the Pogallo River. The San Bernardino River leaves the park's southeastern boundary and flows toward Lake Maggiore, one of the main Alpine lakes.

These two valleys are surrounded by sharp-peaked mountains that are difficult to cross, with an average altitude of around 2000 m. The highest peak is Monte Togano (2301 m). The access to the valleys from the lower San Bernardino River valley is arduous. The rivers have deep-set beds and the valleys are very narrow. To sum up, the whole region was extremely isolated and difficult to access. Currently, the only existing paved road entering the park is a very narrow, zigzagging one that reaches the hamlet of Cicogna in the lower Pogallo valley.

Nowadays, the national park is densely forested (Figure 2). However, this tree-laden appearance hides a complex land use history of pastoral, agricultural, and forestry activities. Since prehistoric times, humans intensively exploited the region (Figure 3) but, according to Carlo Tosco (2016), the deforested landscape that was present until the twentieth century originated in the Middle Ages. For centuries, the region consisted of communal pastures used for seasonal and transhumance practices. It is recorded that the last shepherds abandoned these practices in 1969. It is worth mentioning that biodiversity is higher in open landscapes, such as meadows that are linked to livestock, rather than in the new forests (Höchtl 2005, 2006, 2007). Thus, the end of the transhumance system caused a severe loss of biodiversity.

The woods are characterized by broadleaved trees, but presumed local natural species are limited because of ancient human activity. Nowadays, beeches (*Fagus sylvatica*) dominate the landscape, covering around 40 percent of the national park. Under the beech altitudinal range, the woods are dominated by chestnut trees (*Castanea sativa*, almost 15 percent of the protected area), which were introduced by humans (Larcher and Salvatori 2016). Also, the timber

Fig. 2—The current appearance of the landscape: the Pogallo Valley near Cicogna. Picture by Valerià Paül (09/01/2018).

FIG. 3—Prehistoric petroglyphs over the San Bernardino river valley, with Lake Maggiore in the background. Picture by Valerià Paül (09/01/2018).

industry was significant in this area until the 1960s, as timber could be sent downriver to Milan, a practice in use since the Middle Ages. The last phase of intensive timber industry occurred during the first half of the twentieth century, thanks to a system of cableways that crossed the valleys.

The national park is divided into 13 municipalities. Except for three small hamlets (Genestredo, in the municipality of Vogogna; Colloro, in Premosello Chiovenda; and Cicogna, in Cossogno), the settlements of these municipalities are outside the national park. This means that the evolution of the population of these 13 municipalities is not the population of the park itself. According to the census, the depopulation process from 1861 (18,252 inhabitants) to 2019 (12,749 inhabitants) has resulted in a reduction of 30.15 percent—however, this does not coincide with land abandonment as these people live outside the national park. With reference to the three hamlets located inside the national park, two (Genestredo and Colloro) are adjacent to the boundaries. Only Cicogna is in the middle of the park. Its demographics are representative of the human dynamics of the Val Grande: the hamlet had 700 inhabitants in the 1930s and nowadays has 21 inhabitants, an overall reduction of 97 percent.

The idea of designating Val Grande as a protected area began in the 1950s when it was first discussed in parliament. The first strict nature reserve of the Italian Alps, Monte Pedum, was so designated in 1971. During the 1970s and the 1980s, several environmental associations and local stakeholders (politicians, inhabitants, cultural and hiking associations, local reporters, and the like) promoted the idea of

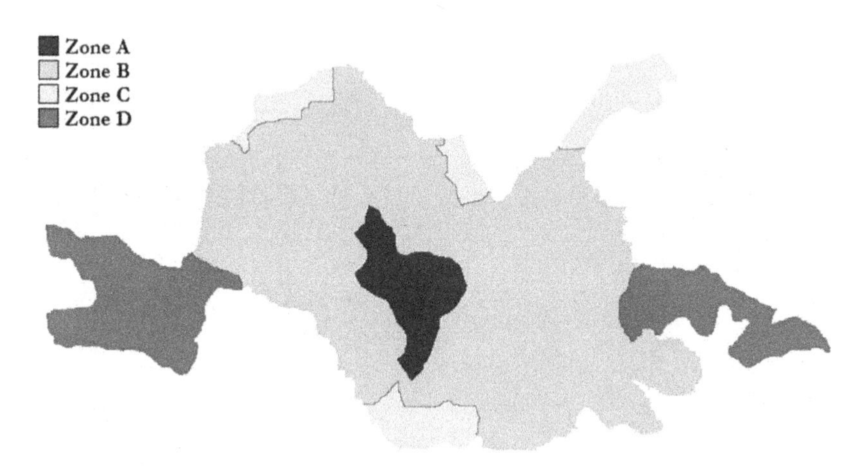

FIG. 4—Simplified Zoning Scheme. Source: Giacomo Zanolin, 2020.

establishing a natural park in Val Grande. Finally, in 1992, the national park was approved according to a Ministerial Decree of 1989 and Act 394/1991. The first management authority was set up in 1993, the first president was nominated in 1994, and the first director in 1995. The current boundaries of the national park were established in 1998, after an enlargement requested by some municipalities.

In accordance with Act 394/1991, a zoning plan was approved for the national park in 1999. Hence, it consists of the following zones (Figure 4):

- A — highest protection. It matches the Monte Pedum strict nature reserve. Access is prohibited, except for specific reasons such as research, environmental monitoring, etc.
- B — in this zone, consisting of the main portion of the National Park, visitors are allowed access, but economic activities are prohibited.
- C — agriculture and livestock are allowed, together with tourism (for example, farm tourism).
- D — includes hamlets and farms, where local economic activities are allowed.

In 2013, the Val Grande National Park was included in the UNESCO Sesia-Valgrande Geopark. In 2018, UNESCO also designated the national park together with other neighboring areas as the Ticino Val Grande Verbano Biosphere Reserve, covering an area of 332,000 hectares. It is considered to be a first step toward the creation of a transboundary biosphere reserve with Switzerland, covering all the Ticino River basin. Since late 2019, a proposal for enlarging the Val Grande National Park has been under discussion.

Methodology

This paper is primarily based on an analysis of texts published about the area. Based on our research goals and following Gordon R. Waitt's (2011) recommendations, we selected a number of publications prioritizing those dedicated only to Val Grande for geographical, environmental, historical, and tourism purposes. After an extensive search, we identified 64 books of this type, published since the 1970s, but mostly after the national park's designation in 1992. After careful review and analysis, we identified about a dozen of these books as the most critically relevant. We extensively refer to this selection in the results section. Most of the other publications pursue the same topics as the main books considered here, as we will highlight when dealing with the second narrative.

The corpus of books has been analyzed by means of content analysis. According to Kothari (2004, 110), this procedure consists of "analysing the contents of documentary materials." José Ignacio Ruiz Olabuénaga (1999, 194–195) states that content analysis is necessarily constructivist given that interpretation becomes central in this method. Accordingly, we may acknowledge that the meaning of the original writer and the meaning of the researcher might not match, and readers of the subsequent work may also infer a third sense. Ruiz Olabuénaga (1999) and Kothari (2004) concur that content analysis may refer to subtle or latent contents, of which the original writer might not be aware, but that the researcher may attempt to capture. This echoes Waitt's (2011) Foucauldian discourse analysis.

According to Rodolphe De Koninck (2001), fieldwork is crucial in geographical research because it allows us to comprehend the specific characteristics of places by means of direct perception. In this respect, Morange and Schmoll (2016) argue that observation becomes essential in the discipline of geography, in line with an anthropological approach. Additionally, fieldwork has also enabled discussions with people in contact with Val Grande via a shared, vivid experience in order to understand its inherent meanings. Those meanings might not be captured through the analysis of documents or through specific methodological instruments such as interviews and focus groups. It is important to mention that one of the authors of this paper walked along almost all of the footpaths in Val Grande National Park over a period of 15 years and frequently used some of the mountain huts that will be analyzed below, echoing the participant observation method as described by Morange and Schmoll (2016). In doing so, the authors were able to witness the evolution of the national park and interpret the effects of the policies implemented in the field.

Results

The analysis we carried out identified three main narratives about wilderness in Val Grande that have converged after the national park's designation in 1992.

The first is quite apparent and derives from the manifest use of the word itself, or variations such as "wild nature," as applied to the area. The second is devoted to the approach developed in local history books. The third deals with Val Grande viewed as a mountaineering attraction. This section is structured following these three interwoven narratives, which are not addressed chronologically but which attempt to make sense of the discursive developments. Subsequently, a fourth section focuses on the national park's institutionalization and formal policies, merging the three previous narratives.

WILDERNESS NARRATIVE

As already noted, in the 1950s and 1960s traditional human activities disappeared from the area. During that time few hikers visited the region. However, there is a meaningful account of one of these experiences, from Teresio Valsesia, who is essential in the development of the wilderness narrative.

"For us hikers, the Val Grande's 'golden years' were those between the 1950s and 1960s. ... These were the years of the discovery of a fulfilling training ground of wildness and freedom. [We] walked for days and days without meeting anyone ...

"Great silence and sunny horizons. Nature that was wholesome and rewarding. Tangled forests and trails engulfed by vegetation. The most common option was to get lost and so to feel adventurously connected with the vegetation." (Valsesia n.d., 1).[1]

In 1964, Mario Pavan, from the University of Pavia, led an expedition of naturalists and forest engineers into the area. He had, in 1959, promoted the designation of the first Italian strict nature reserve of Sasso Fratino in Emilia-Romagna, with the understanding that people had to be banned from certain parts of the Italian territory in order to exclude any human interference with a desired, pristine, natural environment. Interestingly, the expedition was supposed to research the area in a single day, arriving by helicopter in the central part of the Val Grande. However, a sudden storm meant that they could not be picked up that evening and they were forced to stay in the area, in tough conditions, for four days. The development of the notion of wild nature in Val Grande benefits from their personal experiences. The paper reporting their results concluded that the area was "still not yet subjugated or perhaps was even unknown to humans .... Up there, nature was ... intact .... Only some rudimentary stone huts testified to the only human presence some centuries before" (Ingannamorte 1965, 16). The conclusion was in the title of the report: "Val Grande, the new Italian strict nature reserve." Indeed, as already mentioned, Monte Pedum, at the core of Val Grande, was designated as a strict nature reserve in 1971.

In the 1970s, Franco Zunino frequently used the word wilderness in English when referring to Val Grande. The first time he might have said this was at an

FIG. 5—Hard covers of Valsesia's book. Sources: Valsesia (1985, 1992, 2008).

international conference in Johannesburg in 1977, where he claimed that Val Grande had to be preserved as the "first wilderness area in Europe" (Valsesia 1985 [ed. 2006], 51). In 1980, he published a book in Italian entitled *Wilderness—A New Requirement for Nature Area Conservation*, again using the English word. According to Valsesia (1985 [ed. 2006], 51), Zunino openly declared that Val Grande was "an area with wilderness values that must be preserved at any cost ... It is a singularity, an island that has survived the pressures of civilization.".

A milestone in the discursive construction of the wilderness notion in Val Grande is Teresio Valsesia's (1985) book, significantly entitled in Italian *Val Grande—The Last Paradise*. The subtitle declared that it was *The Largest Wild Area of Italy*. This book, widely circulated, contained chapters devoted to natural and environmental features (for example, flora and fauna), and others to human intervention (for example, forestry, chestnut plantations, and toponymy). Valsesia was aware that there had been intense human presence and wanted to show what it was like for people existing in an environment that he understood to be extremely difficult to live in. At the same time, from his walks since the 1950, he had realized that there was no one living, working, or experiencing nature in Val Grande, triggering in him a deep sensation of the rising levels of wilderness. Since the designation of the national park in 1992, the book has had several new editions, with subtle changes in the subtitle: at first, mentioning the *Wildest National Park* (1993), and later *National Park* only (2006) (Figure 5). The word *wild* has disappeared, although the changing picture used on the hardcover shows that the wilderness notion remains at the forefront with a spectacular image of the forests.

LOCAL HISTORY NARRATIVE

Valsesia's (1985) book contained some considerations on local history. Later, local historians devoted complete books to this topic. An example is Nino Chiovini, who interviewed elderly people from the region to record their memories. His main book is dedicated to the area's historical activities, such as shepherding, smuggling, poaching, and woodcutting. (Chiovini 1991 [ed. 2002]). However, he has also written important works recounting his memoirs as a partisan in the area during World War II, recently expanded to include other people's accounts (Chiovini 1966, 1974 [ed. 2005]).

Chiovini is one of the authors of this narrative, but there are many others who have contributed in this direction. There are recurrent themes always linked to the difficulties of working and living in an extreme environment for humans. We find an intertextual dialogue between these works and other repeated stories that, despite hardly mentioning any sources, might have been adapted from previous books, namely Chiovini's. A clear example of this is the obvious correspondence between Chiovini (1991 [ed. 2002], 40–41), Valsesia (1985 [ed. 2006], 133), and Primatesta (2010, 19) that refer to the same early twentieth century anecdote. A shepherd boy had lived since birth in a remote hut in Val Grande for six years. When the time came to leave this place to attend school, he saw, for the first time, a horse (or a donkey, depending on the source) pulling a cart, and he mistook it for a cow with a spinning wheel. Interestingly, the quotes of the words allegedly used by the child are written in the dialect spoken in the region, but with different spelling, respectively: "*vaca muta ch'la tirava 'n grös finarèl*," "*vaca müta ch'la tirava 'n gròss finarell*," and "*vaca muta tirà un grôs finarèl*."

An important part of these books lies in their images rather than texts, in particular in the recovery of old photographs. These show people, mainly men, working, conducting their activities, mostly shepherding, woodcutting, and hunting/fishing. Quite surprisingly, the implemented infrastructure for the timber industry was sophisticated, making it clear that the area was intensively exploited.

This set of books has helped result in the recovery of the locals' lost memory. Importantly, local self-esteem has improved. A case in point is Alberto Paleari, a well-known local alpinist who has written mountaineering guides and novels set in a neighboring alpine region. However, though he was born nearby, he admits that he had never thought about Val Grande as a setting for his writings until 2018.

"A long time ago, in the late 1960s ... nobody knew that there was a Val Grande .... Forgetting the Val Grande for our parents meant forgetting where we came from, forgetting the hardship, the 'lives' they led, the shortages, the need to save, forgetting the cold" (Paleari 2018, 3).

MOUNTAINEERING NARRATIVE

The most prominent person to scale Val Grande is the free climber Ivan Guerini. He was the first to open up several climbing routes in Val di Mello (central Alps, Lombardy) in the 1960s, but, because he considered the area increasingly under pressure from tourists and other climbers, he moved to Val Grande in 1972 (Guerini 1999, 2012). Over the next few years he pioneered several climbing routes, even in Monte Pedum's strict nature reserve. Interestingly, in 1986 he decided to leave the Val Grande mountains as he saw, yet again, that the increasing number of visitors was irreparably changing the region (Guerini 1999, 2012). Guerini's books (written many years later, in 1999 and in 2012) are a unique testimony of his experiences over the years before the Val Grande National Park's foundation. He describes the valley by expressing isolation and loneliness, at a crucial moment in history, in the period of total abandonment.

"The difficulty of finding one's bearings due to the thick foliage boosted the ability to follow a certain direction without references. ... The pace slowed by the tangle of vegetation made it possible to sharpen the capacity to observe which is often inhibited by haste ... and the risky inclines of the steepest slopes strengthened concentration, eroded by unexpected obstacles ... . So the obstacles, more than a challenge to personal limits, turned out to be expressions of landscape elements" (Guerini 2012, 22).

But the fact that Guerini left Val Grande in 1986 demonstrates that the area was becoming a preferred destination for growing numbers of mountaineers. In fact, Chiovini also wrote in 1987 that "the area is always more crowded," with some "negative behaviors" requiring "control and regulations" (Chiovini 1991 [ed. 2002], 63). These mountaineers made use of two essential features from the past that had been abandoned: the trails and the huts. Valsesia's (1985) book mapped some of these trails but, by then, they were neither marked and nor evident in the landscape. Some old huts initially were used for sleeping, but sometimes work was carried out by the mountaineers to adapt the buildings for accommodation (Figure 6).

Mountaineers frequently lost their way and, during the 1980s, there were several reported cases of mountaineers who went missing. This contributed to the Val Grande's growing reputation as a dangerous, even bad, place that isstill heeded by mountaineers. Whilst the numbers of mountaineers increased, the published experiences acknowledge a rampant contradiction between the sense of wilderness that they were attracted to and their feelings while walking.

"June 1995 ...

"— Wilderness! — I was swearing to myself, yeah, wilderness, wasn't that what I was looking for?

"The trail, which was indicated on the map, was invisible. [...]

Fig. 6—Alpe Borgo delle Valli. Example of a hut adapted by mountaineers. Picture by Giacomo Zanolin (07/26/2008).

"But where is this Borgo [a hut]? Finally, I emerged into a clearing where I found a trace and, a little later, here was a surprise: in front of the huts, clear in the twilight, burned a crackling fire.

"Gnomes? Elves? Wizards?

"No.

"Men, fishermen, brothers." (Bellavite 2006, 32–33).

THE RECONFIGURATION OF THE NARRATIVES BY THE NATIONAL PARK

The new Val Grande institution held a conference in 1996. This was mainly devoted to discussing the compatibility between wilderness protection (*wilderness*, in English, was used in the title of the conference in Italian) and the development of tourism. Regarding wilderness conceptualization, one of the keynote speakers recognized that the area was "not necessarily a virgin forest —now practically unknown in Europe—[,] it rather means letting nature take its course, 'abandoning the exploitation'" (Gotz 1996, n.p.). This was a recognition that wilderness had to be especially nuanced. Roberto Gambino (2016) has qualified the unique conceptualization of wilderness in Val Grande since the national park's early years as *wilderness di ritorno* ("returning wilderness"), now that human exploitation ceased.

At that conference, Franca Olmi, the first president of the national park, declared that tourism was pivotal to the park. Tourism was planned to be based

FIG. 7—Alpe Pian di Boit. Example of a hut restored by the National Park to experience wilderness in Val Grande. Picture by Giacomo Zanolin (6/10/2007).

on the notion of discovery and exploration of the wilderness, making use of the eco-tourism approach that was intensively developed in the 1990s, and also mentioning that wilderness was an opportunity to experience solitude and to move away from civilization through immersive experiences in nature. In this sense, she stated that "we should ask ourselves which type of tourism can be expected in Val Grande without compromising its uniqueness as a wild area" (Olmi 1996, n.p.). First and foremost, she acknowledged the natural features of the area, but also "various historical and cultural features" (Olmi 1996, n.p.). Ultimately, these developments could imply a socioeconomic development of the protected area.

Since its inception, the main tourism strategy developed by the national park has been to clean up several old trails and old huts to form part of a proper infrastructure for hikers in zone B. These huts are something of an oxymoron: through them the aim is to live an experience of wild nature, but sleeping there is the tangible proof of long-term human presence in the valleys. Twenty-seven of these ancient huts have been restored, and have a modest level of comfort, commonly including wooden tables with benches, a wooden chest, a wooden loft space for sleeping, a wood stove or a fireplace, and a small solar panel to produce power to light the building (Figure 7).

The bid for tourism of the national park has gained momentum in recent years, especially since the park joined the ECST in 2013. The most popular intervention has been the implementation of a zip- line, which, thanks to gravity, allows you to "fly" (at more than 120 km/h) over the valley hooked onto a rope, offering a new way of enjoying the forest landscape. This zip line is strictly outside the park boundaries, but has been promoted by the national park itself. A third project under the ECST is the creation of the Val Grande Literary Park, entitled Nino Chiovini.

Discussion and Conclusions

As mentioned above, the aim of this paper is to consider the wilderness notion by means of the Val Grande case study, linking it to the sustainability concept. An initial finding from our analysis is that wilderness cannot be taken for granted. Researchers such as Höchtl (2005, 2006, 2007) had assumed, uncritically, that the Val Grande National Park area was a wilderness region. However, there is enough evidence to argue that we should be cautious about using this category without proper critical interrogation. In this sense, this case study confirms that wilderness should not be seen as a human-free environment, as this is unattainable (Sauer 1925; Dudley 2011). This is relevant for protected areas' planning and management, also in terms of sustainable development, in accordance with Ojeda (2006), and in terms of sustainable tourism, in accordance with Hughes and others (2015).

An outstanding contributor to the dissemination of the wilderness concept in Val Grande is Zunino. He made use of the word in English even when he was writing in Italian—a bizarre use that has persisted across the years. He assumed that the region was truly wild in the sense that human presence was nonexistent. This is in line with the ideological perspectives of deep ecology (Naess 1973; Capra 1995) and the new environmental paradigm (Dunlap and Van Liere 1978; Van Den Born 2001). Thus, biocentric ethics explain why Val Grande was initially conceptualized as wilderness. Beyond Zunino, the narrative of wilderness in Val Grande results from the works of many other authors.

Valsesia is particularly significant because his book has enjoyed wide circulated. Interestingly, his notion of wilderness is more nuanced than Zunino's by linking with the two other narratives set forth in this paper, in particular the local history narrative developed by Chiovini, amongst others. In fact, Chiovini echoes Jacoby's (2001) perspective on the "hidden history" of conservationism. In short, Valsesia's vision is conservationist and ecocentric according to Depraz (2008) and Andreozzi (2017). As shown in Figure 5, the word *wild* in Italian has disappeared from the hardcover title. This may be interpreted in the need to align the book with the national park imaginary and policy making since it was founded in 1992. It has progressively reflected a more commodified vision of wilderness where authenticity is not important, but the capacity to sell a product is (Lovell and Bull 2018).

Guerini's perception has also been relevant. He assumed, like Zunino, that previous human presence was imperceptible. Although Guerini's free-climbing practices were different from traditional mountaineering, the Romantic movement, which valued particular emotional connections with nature (Martínez de Pisón and Álvaro 2002; De Rossi 2014), echoes his words. In this sense, Guerini holds a Romantic gaze in Urry's (2002) terms. When, in 1986, he perceived that Val Grande was beginning to be the object of the collective gaze in Urry's (2002) terms, Guerini abandoned the area. This is a fundamental point of the literature on sustainable tourism: visitor thresholds are critical and can be different on an environmental, social, and even personal level (Hunter 2002; Frost and Hall 2009; Hughes and others 2015).

Moreover, wilderness in Val Grande is also the consequence of the presence of forest engineers, biologists, and other academics such as Pavan. Although Pavan did not use the word *wilderness*, he inspired this notion in future iterations. Interestingly, his colleagues in this campaign (for example, Ingannamorte 1965) wrote that the human presence had dissipated some centuries before, when in reality it had been very active until the previous decade. This is in accordance with Haraway's (1988) denunciation of the Western construction of culture and nature as ontological opposites. Importantly, the wilderness narrative creates a social imagery—in Debarbieux's (2019) terms—free of humans, causing a separation with the communities that had used this space. As this paper explains, the acknowledgment of historical human presence by means of the local history narrative has led to the recovery of self-esteem for these inhabitants. This is central when developing a protected area that cannot be separated from the local inhabitants. In addition, self-esteem is essential in development terms, as a community cannot develop without an awareness of its own identity and its heritage (Ojeda 2006), an important point to take into account when dealing with sustainable development.

Pavan is also essential in the discussion of the regulatory institutionalization of wilderness in Val Grande. First, the strict nature reserve in 1972 resulted from his personal involvement, as did, indirectly, the national park in 1992, although under different circumstances. Hence, Val Grande is one of the "transfers" from the U.S. designations to Europe. However, as pointed out by Nash (1967 [ed. 2014], 1970), Tuan (1996), Depraz (2008), Frost and Hall (2009), and Debarbieux (2019), U.S. wilderness designations are specific to their particular context, so its transferability to Europe, as confirmed in this paper, remains highly problematic. Additionally, strict nature reserves can be correlated with the "Great Wilderness Debate" as coined by Callicott and Nelson (1998) and Nelson and Callicott (2008), in particular the perceived need to set aside areas safeguarding them from human interference. The evidence reported here is strong enough to state that policy makers should think twice before equating protected areas with strict nature reserves.

When the national park was designated, the concept of wilderness had to be reformulated in line with Valsesia's works. Accordingly, the national park's early years were based on three policies that can be linked with the three narratives: protection (linked with wilderness preservation), knowledge (implying an in-depth examination of the natural and cultural heritage of the area, critically including the historical human presence), and promotion. Arguably, this third narrative of tourism has gained momentum, as acknowledged in Franca Olmi's writings. However, the national park has not abandoned the wilderness concept—with an ongoing use of the word in English—given that it constitutes a pivotal identity brand that creates a distinct image for the area as a tourist destination. In terms of sustainability, it seems dubious that tourism must be such an important component of a protected area. In this sense, we maintain that no tourism can be possible if the aim is to preserve unsullied nature, nor can any sustainability or development be possible if humans are excluded. In fact, it may be inferred that Val Grande demonstrates how wilderness can become a mere brand or slogan, even a metaphor, ready-made for tourism purposes, but with few connections with the real history of the territory.

A case in point in the wilderness reconceptualization encountered in the case study is the development of Gambino's "wilderness *di ritorno*" notion. The park subtly assumed that the notion of wilderness as advocated by preservationism and deep ecology (Naess 1973; Capra 1995; Depraz 2008; Andreozzi 2017) is unattainable and coined a new concept adapted to the particular circumstances of the region echoing the livelihood approach regarding secondary forests, argued by Hecht (2004) and linked to ideas of territoriality, identity, and local knowledge systems. Franco Brevini (2013) has already warned that the wilderness concept is continually revisited and "di ritorno" seems to be a new variation. This is a worthwhile addition to the existing literature on wilderness. The analysis carried out here has been in line with Noel Castree's (2001) call for a critical questioning of our knowledge of nature (and, consistently, wilderness) (Whatmore 2002), and its sustainability implications. In doing so, this paper reinforces the idea that wilderness, like nature, has to be understood as a social construction, corroborating Castree (2001; 2014), Moeckli and Braun (2001), and Demeritt (2002). This is important for our understanding of national parks across the globe, enabling us to relate to them with greater awareness of the value of the heritage they protect, a prerequisite if we are to adopt a responsible and sustainable attitude toward them. On this basis, further research might focus specifically on sustainable tourism as conceived in Val Grande so as to question the park administrators' default preference for tourism.

By focusing on a specific case study, this paper has highlighted some paradoxes referring to wilderness and sustainability, calling attention to the pivotal role achieved by tourism in the practical developments of both these concepts in Val

Grande. Further research should be concentrated on other comparable case studies where sustainable tourism is developing in protected areas, for instance by means of the ECST, to corroborate if the concept of wilderness always plays the same role.

The concept of wilderness has been and continues to be used and played upon as a socio-cultural-political construction in the case of Val Grande and, through these three narratives, demonstrates that there has been a shift between different cultural constructions of the concept, with different policy implications, namely from a more strict conservationist approach to a more utilitarian one. The latter aims to link the concept of wilderness to tourist practices and to the tourism potential of the area.

Note

[1] This and the following are our translations.

ORCID

Giacomo Zanolin http://orcid.org/0000-0002-5059-5618

Valerià Paül http://orcid.org/0000-0003-3007-1523

References

Andreozzi, M. 2017. *Biocentrismo Ed Ecocentrismo a Confronto. Verso Una Teoria Non-antropocentrica Del Valore Intrinseco*. Milano: LED.

Bellavite, P. 2006. *La mia Valgrande. Sentieri e pensieri nella natura selvaggia del Parco Nazionale*. Comune di Beura Cardezza e Comune di Cossogno.

Blewitt, J. 2008. *Understanding Sustainable Development*. London: Earthscan.

Borowy, I. 2018. Sustainable Development and the United Nations. In *Routledge Handbook of the History of Sustainability*, edited by J. L. Caradonna, 151–163. London: Routledge.

Brevini, F. 2013. *L'invenzione della natura selvaggia: Storia di un'idea dal XVIII secolo a oggi*. Torino: Bollati Boringhieri.

Callicott, J. B. 1998. The Wilderness Idea Revisited. The Sustainable Development Alternative. In *The Great New Wilderness Debate*, edited by J. B. Callicott and M. P. Nelson, 337–366. Athens: The University of Georgia Press.

Callicott, J. B., and M. P. Nelson, eds. 1998. *The Great New Wilderness Debate*. Athens: The University of Georgia Press.

Capra, F. 1995. Deep Ecology. A New Paradigm. In *Deep Ecology for the Twenty-first Century*, edited by G. Sessions, 19–25. Boston: Shambhala.

Cassatella, C., ed. 2016. *Dal paesaggio della sussistenza a quello della wilderness*. Parco Nazionale Val Grande.

Castree, N. 2001. Socializing Nature: Theory, Practice, and Politics. In *Social Nature. Theory, Practice, and Politics*, edited by N. Castree and B. Braun, 1–19. Malden: Blackwell.

———. 2014. *Making Sense of Nature. Representation, Politics and Democracy*. London: Routledge.

Chiovini, N. 1966. *Verbano giugno '44*. Verbania, Italy: Comitato della Resistenza di Verbania

———. 1974. [ed. 2005]. *I giorni della semina*. Verbania, Italy: Tararà.

———. 1991. [ed. 2002]. *Mal Di Valgrande*. Verbania, Italy: Tararà.

De Koninck, R. 2001. Du terrain à l'amphi. Le mandat des géographes. In *Géographie et societé. Vers une géographie citoyenne*, edited by S. Laurin, J.-L. Klein, and C. Tardif, 123–137. Sainte-Foy: Presses de l'Université du Québec.

De Rossi, A. 2014. *La costruzione delle Alpi. Immagini e scenari del pittoresco alpino (1773–1914)*. Rome: Donzelli.

Debarbieux, B. 2019. *Social Imaginaries of Space. Concepts and Cases*. Cheltenham: Edward Elgar.

Demeritt, D. 2002. What Is the 'Social Construction of Nature'? A Typology and Sympathetic Critique. *Progress in Human Geography* 26(6):767–790. 10.1191/0309132502ph402oa

Depraz, S. 2008. *Géographie des espaces naturels protégés. Genèse, principes et enjeux territoriaux.* Paris: Armand Colin.

Descola, P. 2015. *Par-delà nature et culture.* Paris: Gallimard.

Dudley, N. 2011. *Authenticity in Nature. Making Choices about the Naturalness of Ecosystems.* London: Earthscan.

Dunlap, R. E., and K. D. Van Liere. 1978. The 'New Environmental Paradigm.'. *The Journal of Environmental Education* 40(1):19–28. 10.3200/JOEE.40.1.19-28

Frost, W., and C. M. Hall. 2009. *Tourism and National Parks. International Perspectives on Development, Histories and Change.* London: Routledge.

Gambino, R. 2016. La Val Grande tra wilderness, Parco ed Ecomuseo. In *Dal paesaggio della sussistenza a quello della wilderness*, edited by C. Cassatella, 13–19. Parco Nazionale Val Grande.

Gotz, A. 1996. La wilderness nelle Alpi. In *Atti del Convegno Wilderness e turismo integrato Opportunità o conflittualità?* http://www.parks.it/parco.nazionale.valgrande/documenti/wilderness-turismo-integrato/19ottobre96/04.html.

Guerini, I. 1999. *Val Grande. Mondo segreto di rocce e piante.* Verbania: Alberti Libraio editore.

Guerini, I. 2012. *Val Grande. Storia Esplorativa Dei Territori Sconosciuti.* Lecco: Alpine Studio.

Haraway, D. 1988. Situated Knowledges: The Science Question in Feminism and the Privilege of Partial Perspective. *Feminist Studies* 14(3):575–599. 10.2307/3178066

Hecht, S. B. 2004. Invisible Forests: The Political Ecology of Forest Resurgence in El Salvador. In *Liberation Ecologies. Environment, Development, Social Movements*, edited by R. Peet and M. Watts, 58–94. London: Routledge.

Höchtl, F., and others. 2005. "Wilderness": What It Means When It Becomes a Reality—A Case Study from the Southwestern Alps. *Landscape and Urban Planning* 70:85–95. 10.1016/j.landurbplan.2003.10.006

———. 2006. Pure Theory or Useful Tool? Experiences with Transdisciplinarity in the Piedmont Alps. *Environmental Science & Policy* 9:322–329. 10.1016/j.envsci.2006.01.003

———. 2007. Building Bridges, Crossing Borders: Integrative Approaches to Rural Landscape Management in Europe. *Norsk Geografisk Tidsskrift-Norwegian Journal of Geography* 61:157–169. 10.1080/00291950701709150

Hughes, M., and others, eds. 2015. *The Practice of Sustainable Tourism. Resolving the Paradox.* London: Routledge.

Hunter, C. 2002. Aspects of the Sustainable Tourism Debate from a Natural Resources Perspective. In *Sustainable Tourism: A Global Perspective*, edited by R. Harris, and others, 3–23. Oxford: Elsevier.

Ingannamorte, F. 1965. Avventure nella Val Grande, la nuova riserva naturale integrale italiana. *Notiziario forestale e montano* 135(10):13–21.

Jacoby, K. 2001. *Crimes against Nature. Squatters, Poachers, Thieves, and the Hidden History of American Conservation.* Berkeley: University of California Press.

Kothari, C. R. 2004. *Research Methodology. Methods and Techniques.* New Delhi: New Age International Publishers.

Larcher, F., and L. Salvatori. 2016. I paesaggi agroforestali: Struttura, qualità e dinamiche". In *Dal paesaggio della sussistenza a quello della wilderness*, edited by C. Cassatella, 133–155. Verbania, Italy: Parco Nazionale Val Grande.

Lovell, J., and C. Bull. 2018. *Authentic and Inauthentic Places in Tourism. From Heritage Sites to Theme Parks.* London: Routledge.

Martínez de Pisón, E., and S. Álvaro. 2002. *El sentimiento de la montaña. Doscientos años de soledad.* Madrid: Desnivel.

Meli, F. 2007. *La letteratura del luogo.* Milan: Arcipelago.

Moeckli, J., and B. Braun. 2001. Gendered Natures: Feminism, Politics, and Social Nature. In *Social Nature. Theory, Practice, and Politics*, edited by N. Castree and B. Braun, 112–132. Malden, Mass.: Blackwell.

Morange, M., and C. Schmoll. 2016. *Les outils qualitatifs en géographie. Méthodes et applications.* Paris: Armand Colin.

Naess, A. 1973. The Shallow and the Deep. Long Range Ecology Movement. A Summary. *Inquiry* 16:95–100. 10.1080/00201747308601682

Naredo, J. M. 2007. Crecimiento insostenible, desarrollo sostenible. In *Geografía humana. Procesos, riesgos e incertidumbres en un mundo globalizado*, edited by J. Romero, 421–476. Barcelona: Ariel.
Nash, R. F. 1967. [ed. 2014]. *Wilderness and the American Mind*. New Haven: Yale University Press.
———. 1970. The American Invention of National Parks. *American Quarterly* 22(3):726–735. 10.2307/2711623
Nelson, M. P., and J. B. Callicott, eds. 2008. *The Wilderness Debate Rages On*. Athens: The University of Georgia Press.
Nijkamp, P. 1990. *Regional Sustainable Development and Natural Resource Use*. Washington D.C: World Bank Annual Conference on Development Economics.
Nogué, J., ed. 2018. *Yi-Fu Tuan. El Arte De La Geografía*. Barcelona: Icaria.
Oelschlaeger, M. 1991. *The Idea of Wilderness. From Prehistory to the Age of Ecology*. New Haven, Conn: Yale University Press.
Ojeda, J. F. 2006. Paseando por paisajes de Doñana de la mano de algunos de sus creadores contemporáneos. In *Doñana en la cultura contemporánea*, edited by J. F. Ojeda and others, 171–204. Madrid: Ministerio de Medio Ambiente.
Olmi, F. 1996. Val Grande: Un'ipotesi di valorizzazione. In *Atti del Convegno Wilderness e turismo integrato Opportunità o conflittualità?* http://www.parks.it/parco.nazionale.valgrande/documenti/wilderness-turismo-integrato/19ottobre96/01.html.
Paleari, A. 2018. *L'altro lato del paradiso. Cinquant'anni in Valgrande*. Milano: Hoepli.
Primatesta, A. 2010. *La Valgrande Di Ieri*. Domodossola: Grossi.
Ruiz Olabuénaga, J. I. 1999. *Metodología de la investigación cualitativa*. Bilbao: Universidad de Deusto.
Salsa, A. 2009. *Il tramonto delle identità tradizionali. Spaesamento e disagio esistenziale nelle Alpi*. Torino: Priuli & Verlucca.
Sauer, C. O. 1925. The Morphology of Landscape. *University of California Publications in Geography* 2(2):19–53.
Short, B. 2006. Idyllic Ruralities. In *Handbook of Rural Studies*, edited by P. Cloke, T. Marsden, and P. Mooney, 133–148. London: SAGE.
Smith, N. 2007. Nature as Accumulation Strategy. *Socialist Register* 43:16–36.
Tosco, C. 2016. La Val Grande dal popolamento alla Wilderness: Un percorso storico. In *Dal paesaggio della sussistenza a quello della wilderness*, edited by C. Cassatella, 85–93. Verbania, Italy: Parco Nazionale Val Grande.
Tuan, Y.-T. 1996. *Cosmos and Hearth: A Cosmopolite's Viewpoint*. Minneapolis: University of Minnesota Press.
Urry, J. 2002. *The Tourist Gaze*. London: SAGE.
Valsesia, T. 1985. [ed. 2006]. *Val Grande ultimo paradiso. Parco Nazionale*. Verbania: Alberti Libraio editore.
Valsesia, T. n.d. Il Pedum: da montagna inesistente a prototipo della tutela alpina. http://www.parcovalgrande.it/pdf/Relazione_Teresio_Valsesia.pdf.
Van Den Born, R. J. G., and others. 2001. The New Biophilia: An Exploration of Visions of Nature in Western Countries. *Environmental Conservation* 28(1):65–75.
Waitt, G. 2011. Doing Foucauldian Discourse Analysis—Revealing Social Realities. In *Qualitative Research Methods in Human Geography*, edited by I. Hay, 217–240. Don Mills: Oxford University Press.
Whatmore, S. 2002. *Hybrid Geographies. Natures Cultures Spaces*. London: Sage.
Woods, M. 2011. *Rural*. London: Routledge.
World Commission on Environment and Development. 1987. *Our Common Future*. Oxford, U.K.: Oxford University Press.

THE EMERGING MOUNTAIN IMAGINARY OF THE GALICIAN HIGHLANDS: A NEW NATIONAL LANDSCAPE IN AN ERA OF GLOBALIZATION?

VALERIÀ PAÜL and JUAN-M. TRILLO-SANTAMARÍA

ABSTRACT. Mountains have been the object of an intense elaboration of national imaginaries. There is a widespread perception of Galicia as a rural and coastal country based on agricultural and seaside landscapes, with mountains being largely ignored. However, a production of local narratives around the Trevinca Massif has recently taken place, which has become quite widespread, with an emerging mountain landscape imaginary of its own around the notion of the Galician Highlands. This paper discusses these developments in the context of both the changing imaginaries of mountains in Western cultures and the different Galician landscape imaginaries. The research was carried out by means of semistructured interviewing, leading toward obtaining three narratives elaborated from semiotic clustering. These results allow us to infer that global forces, in particular tourism and promotion, have been essential for explaining the emergence and spread of the Galician Highlands as a socio-spatial imaginary. Another conclusion is the relevance of the disputed conception of "natural borders" coinciding with mountains in the case-study area.

Certain mountains hold a deep national meaning for the respective countries they are located in (Nogué and Vicente 2004; Debarbieux and Rudaz 2010; Bernbaum and Price 2013; Della Dora 2016; Herb 2018). A case in point is Japan, quintessentially linked with Mount Fuji, customarily shown with sea waves, cherry trees, and traditional architecture, but nowadays these icons have been replaced by other representations, such as manga and even the shinkansen. In the current era of globalization, these imaginaries are quite commonly projected abroad, attaining worldwide significance. Bernard Debarbieux (2004) proposed a framework regarding spatial imaginaries attached to mountains in Western countries[1] that was broadened, gaining a planetwide dimension, by Debarbieux and Gilles Rudaz (2010). In this model, the national imaginary of mountains is commonly expected. However, other imaginaries are also identified, including global environmentalism and internationalism, which have gained momentum throughout the twentieth century. By examining a case study where the mountain imaginary is contemporarily emerging, we can discover if the national aspect remains significant or assert that, currently, the social construction of mountains bears witness to other dimensions.

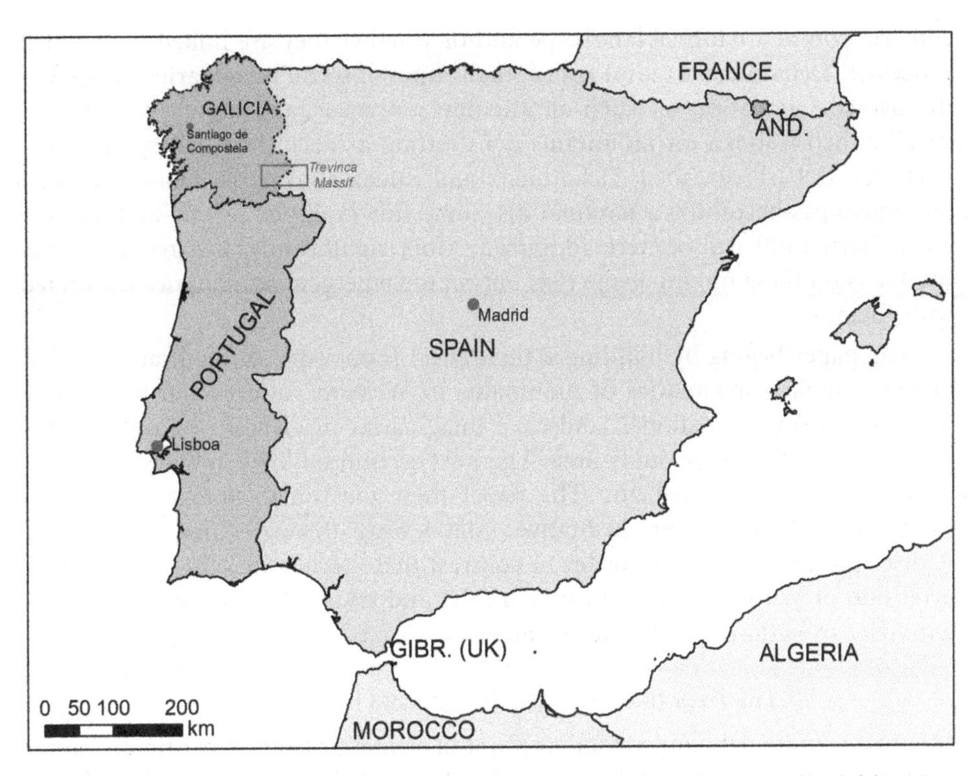

FIG. 1—Location map of Galicia. Source: Base map from https://www.ign.es/, authors' elaboration.

The territory in focus is Galicia (Figure 1), whose national character is based on its own language, culture, and sense of belonging, resulting in a strong identity. Galicia is constitutionally considered a nationality, euphemism for nation, which has experienced devolution within Spain since the late 1970s. As outlined below, mountain landscapes in Galicia have been largely ignored and the national landscape imaginaries have been constructed with other geographical bases in mind. However, throughout the first two decades of the twenty-first century, a new landscape discourse has emerged in the Trevinca Massif (the highest peak of the nation being Trevinca at 2127 m). Obviously, these mountains, the highest ones in Galicia, have always been there, but they have only attained significance in the socio-spatial discourses about Galicia since the 2000s, approximately. In this process of emergent mountain imaginary, the development and extension of the category of the "Galician Highlands" (*Terras Altas de Galicia* in Galician) has certainly been a milestone. The studied region was indeed the "objective" Galician highlands—with a significant land area of over 1,800 m—but no one had used this concept before.

Accordingly, the recent local narratives about mountains, developed in the Trevinca Massif, are analyzed to discern whether these respond to the construction of a national landscape and/or whether they are linked with global processes. Debarbieux's (2019) conceptualization of social imaginaries of space is the overall foundation for such an attempt; moreover, and more precisely, the specific theorizations on mountains constituting a socially framed spatial category (see Debarbieux 2004; Debarbieux and Rudaz 2010; Della Dora 2016) and on landscapes becoming a national discourse (for example, Nogué and Vicente 2004; Herb 2018) will be here addressed. More significantly, this research does not focus on these mountains per se, but on the emergent imaginaries associated with them.

The paper begins by building a theoretical framework through an overview of the changing imaginaries of mountains in Western cultures and by revising the construction of national landscape imaginaries in Galicia, followed by the presentation of the case-study area. The next section outlines how the empirical analysis has been carried out. The paper then goes on to analyze the spatial narratives obtained from qualitative interviewing through analytical coding. A discussion is provided in order to contrast these results in light of the overall evolution of Western mountain imaginaries and taking into account the different Galician national landscape imaginaries.

The Evolving Western Imaginaries of Mountains

Mountains are widely understood as physical elements that are identifiable in the landscape. However, this discernment is not standardized, as the perception of what a mountain is differs among communities. Some imperceptible hills can be regarded as relevant mountains nearby, while orographic massifs, defined as mountainous by geographers and geologists, might not be conceived as such by the people living within or nearby—for a wider discussion in this respect, see Debarbieux and Rudaz (2010) and Veronica Della Dora (2016). Indeed, classical geographers such as Roderick Peattie stated, in the 1930s, that mountains can only be defined by making use of "the imagination of the people who live within their shadow" (Byers et al. 2013, 2).

Hence, in regards to mountains, the focus here is on the "various conceptions [that] have flourished one after the other, each one being a combination of patterns of perception, visions of the world, projects of knowledge and action" (Debarbieux 2004, 404). These successive imaginaries find their correlate in different social, political, historical, and cultural contexts. They cannot all be read as mutually exclusive, but interwoven, overlapping, and in dialogue with each other, in a process that is not always linear.

The "Invention of the Mountain": Between Enlightenment and Romanticism—and Nationalisms

The critical change in the Western perception of mountains took place in the eighteenth century, particularly in the Alps, moving from being perceived as a diabolic image based on Christian conceptions—see Philippe Joutard (1986), William Cronon (1995), and Della Dora (2016)—to an idyllic and, as such, more sympathetic vision. This process has been coined by Eduardo Martínez de Pisón and Sebastián Álvaro (2002, 43) and Isabelle Sacareau (2003, 3) as "the invention of the mountain," an expression that was already used by Joutard (1986) when specifically referring to "the invention of Mont Blanc." This shift can be explained by the Enlightenment; science was considered central to human existence and, by so doing, there was a need to properly conceptualize, objectivize, and define reality. In this sense, Western academics were keen to know the specific location of the mountains and fully study their features, in an attempt to map the world through systematic explorations.

In parallel to the rational interest in the mountains, a significant emotional and sensitive reconfiguration also took place. This is made evident by, for instance, the increasing number of books published in the eighteenth century devoted to mountains. They were described as magnificent places and as the quintessence of natural beauty. A literary milestone is *La Nouvelle Héloïse* by Jean-Jacques Rousseau (1761), in which the Alps were the "natural sanctuary" of Europe (Martínez de Pisón and Álvaro 2002, 39) under a "moral and aesthetic" categorization (Debarbieux 2004, 400). This book was the origin of the taste for mountains in Europe (Bernbaum and Price 2013).

Horace-Bénédict de Saussure's ascent of Mont Blanc in 1787, preceded one year earlier by Jacques Balmat and Michel Paccard, is pivotal for mountaineering, not only in the Alps but also in the world; motivated by science and adventure, Saussure's feat gave rise to mountaineering (Joutard 1986; Martínez de Pisón and Álvaro 2002). Thanks to this Alpine seminal moment, the Alps have become the "model" and "prototype" for mountains in the world (Debarbieux and Rudaz 2010, 40–42). Alpinism has a very intense history led by Alpine Clubs in many countries, whose members have climbed the outstanding mountains of the world throughout the twentieth century (Martínez de Pisón and Álvaro 2002).

Romanticism contributed with new meanings for mountains in Western cultures (Short 1991), causing a multiplication of artistic representations of mountains, namely the Alps, including different types of texts (such as guidebooks, novels, and poetry), music, paintings, lithography printings, the first photographs, and even souvenirs (for instance, porcelain dishes). This imagery became extremely popular and available to a wider public in the nineteenth century (Martínez de Pisón and Álvaro 2002; Nogué and Vicente 2004; Debarbieux and Rudaz 2010; Della Dora 2016).

Joan Nogué and Joan Vicente (2004) analyze how certain mountainous landscapes have become symbols of nationalisms, including not only well established nation-states (for example, the role attributed to the Alps in Switzerland), but also new independent countries (for example, Mount Kosciuszko and, in general, the Australian Alps for Australia) and stateless nations (such as the Pyrenees for Catalonia). Debarbieux and Rudaz (2010) include the particular perception of mountain people within the national narratives in this wave of nationalisms. These are always politically driven discourses, attached either to negative (savage, violent, belligerent . . .) or positive (tough, courageous, loyal . . .) values. According to Guntram Henrik Herb (2018), nationalism endows these landscapes and the people inhabiting them with such connotations because they are conceived as the origin of the nation, as well as entailing a connection between the allegedly unspoiled nature with the unspoiled national soul.

Mountains were attributed an important role as "natural borders" under the Western nationalist conception of mountains in the nineteenth century (Debarbieux and Rudaz 2010; Fall 2010). The origin of this notion can be traced back to Cardinal Richelieu's attempt to fix the French borders coinciding with the Pyrenees and the Alps in the seventeenth century (Sahlins 1990).

NATURAL AND ENVIRONMENTAL VIEWS ON THE MOUNTAINS

Liberal nation-states have developed nature-protection policies in mountain areas since the nineteenth century (Price and Kohler 2013). The designations of Yellowstone and Yosemite as parks in the second half of that century are widely seen as milestones of this policy. However, other achievements can be noted; consider, for instance, the case of forestry policies, such as compulsory afforestation affecting significant parts of mountain areas in the Alps (Debarbieux and Rudaz 2010; Cunha and Price 2013). These designations evidence that the policies were not really based on environmental motives—as could be contemporarily misunderstood—but largely responded to nationalist and protourism intentions, through a particular construction of the notion of wilderness in the United States (Short 1991; Cronon 1995; Depraz 2008; Debarbieux and Rudaz 2010; Woods 2011; Debarbieux 2019).

In any case, national parks as planning devices were rapidly applied to several mountains across the globe. Indeed, other designations soon emerged in the early twentieth century (natural monuments, forest reserves, and the like), leading to a correlation of mountains and protection (Price and Kohler 2013; Arpin and Cosson 2015). Environmentalism has increasingly shaped mountains, especially since the early 1970s, when the United Nations organized the first major conference on international environmental issues in Stockholm in 1972. This conference marked a turning point in the development of environmental policies. In short, mountains are often conceived as "emblematic landscapes [extracted] from artificialisation and modernization" (Debarbieux 2004, 402).

MODERNIST IMAGINARIES: RESOURCES AND DEVELOPMENT

Throughout the twentieth century, mountains have also been seen as areas enduring specific problems, such as depopulation and economic decline, increasingly earmarked for explicit policies delivered by the respective nation-states.

The modernist paradigm initially arrived through dams built for obtaining hydroelectricity and selective manufacturing development (dependent on the availability of natural resources), commonly attached to infrastructures of access (Debarbieux and Rudaz 2010; Price and Kohler 2013). Tourism was already present in mountains from the nineteenth century, but its systematic use for the purposes of economic growth took place throughout the twentieth century. Nowadays, tourism in mountain areas ranges from Alpinist hikers who value solitude to mass tourism and amenity migrants (Nepal and Chipeniuk 2005; Price and Kohler 2013). In many countries, specific aids for mountain farming have also been set up in order to sustain declining farms in the second half of the twentieth century (Majoral 1997; Pujadas and Font 1998; Debarbieux 2004; Debarbieux and Rudaz 2010).

Since the 1960s, regional and/or local policies, agencies, and initiatives for mountain regions have multiplied and, contrary to previous sectoral approaches, these spatial devices envisage integral mountain development (Majoral 1997; Pujadas and Font 1998; Debarbieux and Rudaz 2010; Price and Kohler 2013). In this sense, many mountains have been the object of an accelerated commodification through trademarks and labels promoting their development since the late twentieth century (Debarbieux and Rudaz 2010). In the case of the European Communities (now the EU), devices such as protected food designations and local action groups under the LEADER initiative,[2] launched in the early 1990s, have repeatedly made use of mountain narratives to sustain their identity, basically in economic terms, with inconclusive results (Debarbieux and Rudaz 2010; McMorran et al. 2015).

THE "INTERNATIONALIZATION OF MOUNTAINS"

Over the past few decades, mountains have increasingly been conceived as meaningful at a planet level, rather than at a country level. In this regard, several global initiatives have been significant since the 1970s, such as the Man and the Biosphere Program under the auspices of UNESCO, leading, in turn, to the celebration of the International Year of Mountains in 2002 (Debarbieux and Rudaz 2010; Price and Kohler 2013). Since the early 2000s, civil society initiatives have envisaged avoiding formal politics, such as the World Mountain Forum at the global level or, at the regional level, the Yellowstone to Yukon Conservation Initiative, founded in 1997 (Debarbieux and Rudaz 2010; Chester 2015).

A specific instance is the EU. In the context of the EU integration (including cross-border initiatives), different forms of regional institutionalization of international mountains have taken place (Debarbieux and Rudaz 2010; Balsiger and Nahrath 2015; Debarbieux et al. 2015). The Alpine Region, the Pyrenees, and the Carpathian Range, to name a few, have been the object of intense construction of transnational mountain devices.

Galician National Landscape Imaginaries

According to Justo Beramendi (2007), Galician nationalism dates back to the mid-nineteenth century, initially by asserting the ethnical identity of Galicia—namely by means of a rebirth of the poetry and prose written in Galician, almost absent in formal literature since the Middle Ages—and the need to politically reframe the Spanish state in order to establish a new political status for Galicia. In the early twentieth century, the national character of Galicia becomes overtly acknowledged and the ideological quest for the political institutionalization of Galicia gains momentum, including autonomist, federal, self-determination, and secessionist claims.

Landscape imaginaries have been present since the early variations of Galician nationalism. Federico López Silvestre (2004, 492) remarks on the relevance of the first time the word "landscape" is written in Galician: it is in the prologue of *Cantares gallegos* ("Galician songs") by Rosalía de Castro (1863), which is widely seen as a seminal book of contemporary Galicia. Hence, "landscape" acts as a cultural and social—progressively, also political—discourse for rooting the Galician identity and sustaining the (national) self-esteem and self-identification of the Galician people. This is done by considering the Galician landscape as being similar to prestigious landscapes, such as the Swiss ones—Galicia becomes, quite often, the "Spanish Switzerland"—and contrasting it with the (alleged) core Spanish landscape represented by Castile. Rosalía de Castro gave birth to a Galician landscape understood as green, beautiful, rural, agricultural, maternal, gentle, and full of farmers, broadly recognized in these terms by Galician society in the following decades (López Sández 2008; Murado 2008).

Helena Miguélez-Carballeira (2014) argues that this canonical imaginary has been manipulated for decades because the perception of the landscape as feminine has gradually been linked with the notions of docility, submissiveness, and passivity. This move entails obvious political correlates that are used, for instance, by Spanish institutions to tamper Rosalía de Castro's original voice into a more manageable vision of what Galicia is meant to be. In any case, the Galician landscape ideal is deeply rooted in Rosalía de Castro's construction, privileging this self-representation of Galicia as a fertile garden.

Miguel Anxo Mato (1998), María López Sández (2008), and Miguel-Anxo Murado (2008) show that other Galician writers and historians had already contended with this dominant Galician landscape since the mid-nineteenth century. This opposed landscape narrative is characterized by countermotives, such as masculinity and wilderness; instead of farmers, it is inhabited by epical warriors and mythical heroes; rather than cows and corn (two archetypical elements of the Galician landscape derived from de Castro's works), the landscape is full of wild flora and fauna. Importantly, this narrative favors the coastal landscapes of Galicia with cliffs and creates a link with other Celtic and Atlantic

European nations. This is a landscape narrative that has had a limited effect compared to the first, and main, one. Nevertheless, it is relevant in order to grasp Galician nationalist landscaping.

Since at least the 1930s, Galicia has begun to be customized from the Spanish national perspective, especially for tourism marketing purposes. This image focuses on the southwestern Galician coastal area, with beaches located on the shore of wide *rías* (inlets), accessible and ready to be visited by tourists, quite often depicted in sunny summers. This is an exogenous landscape vision that is opposed by Galician nationalist cultural elites (Paül 2019). However, it has been highly influential and widely adopted in Galicia as a sort of self-representation, converging and merging with the first landscape narrative. Interestingly, such a vision has had a direct effect on the protected natural areas policy, with the first designations in the 1930s affecting coastal areas and a persisting trend to designate coastal lands as protected (including the only national park designated in Galicia: Atlantic Islands of Galicia). Additionally, the Spanish tourism marketing—as well as the one developed in Galicia since devolution in the 1980s—has repeatedly promoted Galician coastlands as a destination, in contrast to inland areas that were not marketed until the 1990s, by means of the increasing attention paid to rural tourism and the Way of Saint James (Lois et al. 2015; Santos and Trillo-Santamaría 2017).

The mountains located in inland Galicia, revealingly, have been largely blotted out from this genealogy of landscape imaginaries. Surprisingly, neither the Galician nor the Spanish nationalist vision has generated a mountainous landscape imaginary in Galicia, an awkward absence in itself given the relevance of mountain landscapes in Western nations, as suggested above. However, this is not to say that there is no mention of mountains (for instance, in the artistic domain, mainly in literature), but they are particular and unstructured. A case in point is the lack of mountaineering in Galicia, in contrast with Spain as a whole, particularly in Madrid, and with Catalonia and the Basque Country, the other two main stateless nations within contemporary Spain, with a long history of very active mountaineering clubs (Martínez de Pisón and Álvaro 2002), although there are some humble exceptions that will be partially considered in the next section. To sum up, the mountainous landscape is expected in Western nations but Galicia has not followed this pattern until quite recently. This paper will consider when and how this emergence has taken place.

The Trevinca Massif

A vast region of more than 850 km^2 extends around the Trevinca peak without any settlements. The hamlet nearest to the peak, Ponte, is 8 km away as the crow flies, but this is an exception given that most of the hamlets are located more than 15 km from the peak, generating an empty region from a demographical perspective. This is due to the climate, which is extremely cold in winter and the

FIG. 2—The Trevinca peak in the background, seen from the south-east in late winter: most of the pastures are covered by snow. Source: Picture by Valerià Paül (23/3/2016).

region is covered by snow for several months (Figure 2). Inhabitants have only settled at lower altitudes. The highest hamlet in Galicia is Cepedelo (1340 m), more than 20 km from the Trevinca summit; above this altitude, permanent population has been historically unmanageable. The area is extremely sloping, making mobility difficult; the 8 km from Ponte to the peak have an elevational gradient of more than 1000 m.

The municipalities[3] in this massif have been depopulated over the last century: demographics peaked in the 1930 Census, with 30,558 inhabitants, and nowadays the population living therein is 7,721 (2019) (Figure 3). Overall, there has been a 75 percent drop in 89 years. The most dramatic decrease has been experienced in the municipality of A Veiga, given that the population registered in 2019 was only 11 percent of the figure in 1930; indeed, there are currently 889 inhabitants distributed among 29 different hamlets, some of them with only five or seven inhabitants (Figure 3).

From a landscape-history perspective, these lands have been devoted to seasonal pastures at least since the Middle Ages, as is the case with many other mountains (Cunha and Price 2013). During summer, thousands of cows and sheep grazed the mountains, enduring the consequences associated with wolves and other feral animals (Figure 4). During the snowy season, the livestock were sheltered in the hamlets or, by means of transhumance routes, transported hundreds of kilometers away. These activities required burning to keep the landscape clear of trees. These practices disappeared throughout the second half of the

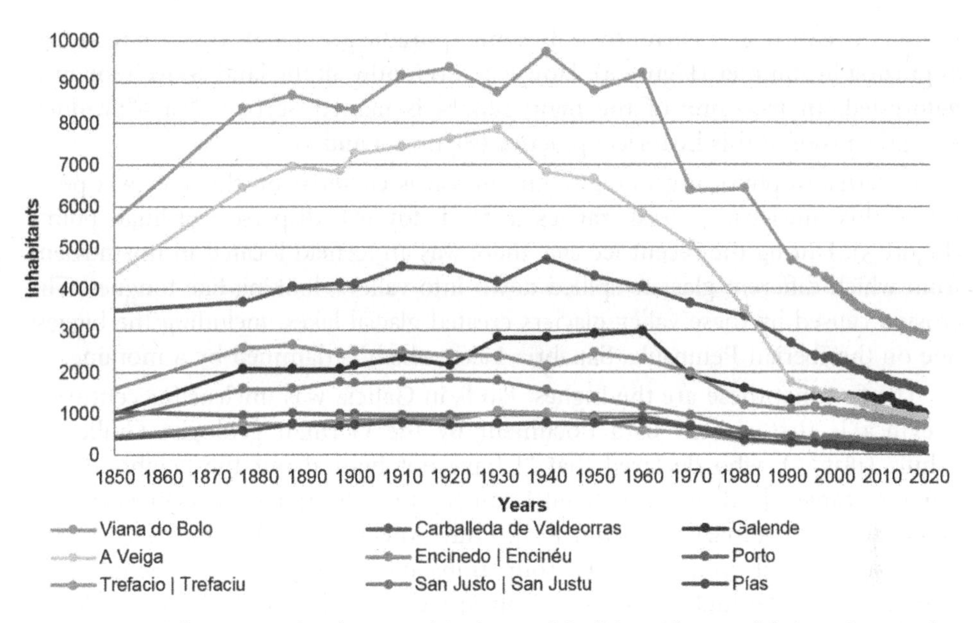

FIG. 3—Population variation (1850-2019) in the municipalities with part of their territory in the Trevinca Massif. Source: www.ine.es, authors' elaboration.

FIG. 4—A high valley (an altitude of 1,600 m) in the middle of the Serra Calva range, with stone walls for fencing in the livestock. Source: Picture by Valerià Paül (22/7/2017).

twentieth century and currently only some sporadic groups of cows are recorded as grazing in summer (Figure 4). However, even now all the land above 1400 m is deforested. In fact, one of the main ranges is named Serra Calva—"Swidden Range"—proof of this historical practice (Figures 4 and 5).

In terms of physical geography, the massif is centered on the Trevinca peak. From this nucleus, several ranges extend toward disperse cardinal points (Figure 5). During the recent ice age, there was an icefield located in this nucleus from which different glaciers spilled down into valleys, looking like tongues. The erosion caused by these valley glaciers created glacial lakes, including the biggest one on the Iberian Peninsula, Sanabria Lake, which is dammed by a moraine.

The fact that these are the highest lands in Galicia was unclear for centuries. Trevinca is first quoted in a document by the German geologist Guillermo Schulz (1835, 7), who declared that "I have not figured out the heights of the different ranges [...] but in general I can say that the ranges in Galicia do not attain an extraordinary elevation [...]. Only Ancares and Trevinca [...] retain some snow until mid-summer" (our translation). Significantly, Schulz was dubious about the highest lands in Galicia (Ancares does not reach 2000 m in height). Another foreigner, also working in mining, was possibly the first to measure the height of Trevinca in the early 1900s—the Belgian engineer Edgar d'Hoore, who managed the rich wolfram mines of the region. He is the first recorded mountaineer in the area and was known by Gonzalo Gurriarán, the founder of the first hiking club in Galicia in 1944 (Gurriarán 2005).

The context in which mountaineering was born in Galicia in general, and in the Trevinca Massif in particular, is ambiguous. The activities began in the 1940s, the first decade of the Franco dictatorship, which repressed Galician culture, language, and identity. Although Gurriarán tried to involve the Galician elites, the hiking club was a modest one. There were continuous plans to open a ski resort, which never materialized (Paül et al. 2019). Some buildings in Fonte da Cova are the result of this pursuit (Figure 6). Additionally, the few Galician cultural activists who were active during Franco's dictatorship were not keen on mountaineering, as the geographer Ramón Otero Pedrayo evidences; a well-known Galician nationalist, he was repeatedly invited to the mountains by Gurriarán in the 1940s–50s but he never went there, possibly because he did not feel that they matched his landscape imaginary, much dependent on de Castro's canon (Paül 2019), as previously noted.

However, and despite the harsh context of Franco's dictatorship, in 1953 Gurriarán led a group to protest against the project to build an artificial dam on the Sanabria Lake moraine that would have destroyed the glacier-born landscape (Gurriarán 2005). Fortunately, the dam was never built and this was the beginning of a process to designate it a protected area. Eventually, the "Sanabria Lake and surroundings" area was designated a natural park in 1978 by the Spanish government, covering part of the municipality of Galende (further extended to a wider area in 1990, and again in 2017; Figure 5).

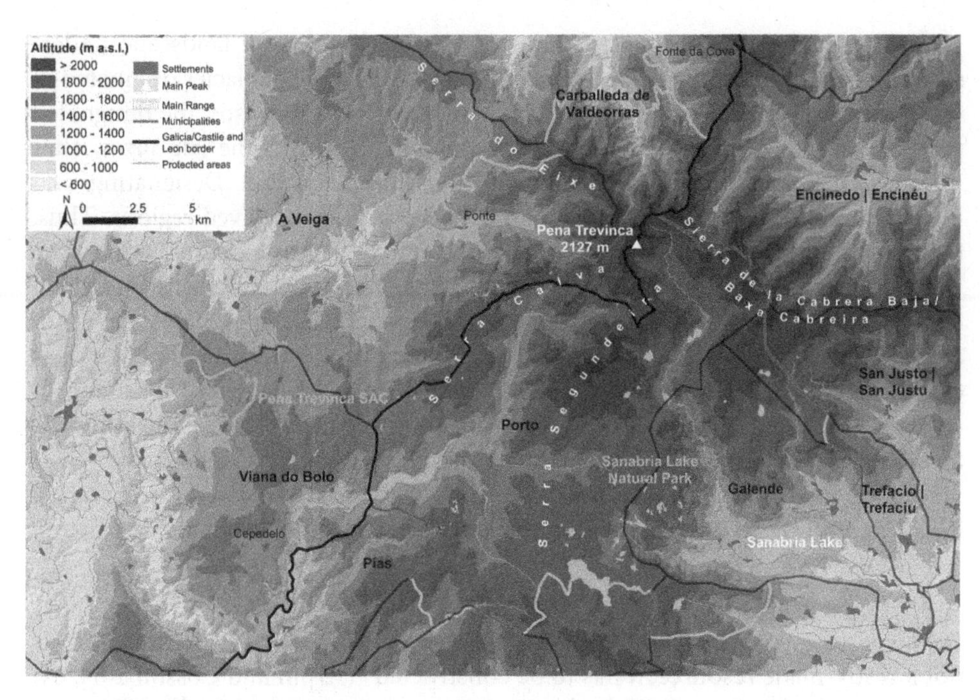

Fig. 5—Regional map showing the toponyms mentioned in the text. Source: Base map from https://www.ign.es/, authors' elaboration.

Fig. 6—The Fonte da Cova hotel built in the 1970s that was supposed to be the center of a ski resort. Source: Picture by Valerià Paül (27/5/2016).

Although the small group of mountaineers' perception of landscape understood Sanabria Lake to be the terminal moraine of the ancient glacier originated in the Trevinca peak—thus part of the Trevinca Massif—in the 1950s, the designated park has been centered on the lake itself since the 1970s. In the late 1970s, after Franco's death, autonomous regions were created in Spain. Designating and managing protected areas became a competence of these devolved regions. Thus, the natural park was transferred to the Castile and Leon Autonomous Region, the Trevinca massif being split between Galician territory (western half) and Leon territory (eastern half) (Figure 5). While in Galicia these mountains have not attracted attention, in the natural park counterpart, tourism development has escalated. The Galician side was unprotected until the EU impelled local authorities to designate areas as part of the Natura 2000 network.[4] Under this initiative, the Galician government designated a Site of Community Importance (SCI) named Trevinca, which was converted into a Special Area of Conservation (SAC) in 2014 (Figure 5), with substantially weaker protection than a natural park.

Methodological Considerations

Following the conceptualization carried out above, "mountain as a category is not a ready-made resource, it has to be constructed" (Arpin and Cosson 2015, 63) as "a social [...] category" (Debarbieux 2004, 404). Rather than looking into this construction by examining materials such as texts or maps (see Debarbieux and Rudaz 2010; Della Dora 2016), our approach favors interviewing. This is because José Ignacio Ruiz Olabuénaga (1999) understands that social constructivism has to be studied by means of this research technique.

Interviewees were selected due to having participated in "attribut[ing]" new "meanings" (Valentine 2005, 111) to the area. Additionally, in these interviews it was essential to comprehend "the processes which operate in particular social contexts" (Valentine 2005, 111) in order to gain a social dimension beyond mere personal considerations. The reader must account for the fact that one of the authors of the paper has been mountaineering in the region, as well as being a member of a local hikers' club, for more than two decades, and the involvement of them both in developing the contents of a visitors' center for the region inaugurated in 2017, as reflecting a bias. On the other hand, this situation has facilitated easy access to several interviewees. Additionally, snowballing has been practiced to recruit other informants (Valentine 2005). It is essential to point out that interviewing does not seek representativeness (Ruiz Olabuénaga 1999; Valentine 2005).

Thirty-six in-person interviews were conducted from February to April 2017; twenty-nine of them are considered meaningful for this paper. Their profiles are challenging as most have various roles, but they can be clustered as follows: seven activists (mountaineers, environmentalists, and such); seven entrepreneurs (mostly in tourism, but not exclusively); six local politicians (five mayors and one deputy mayor); four civil servants working in local governments, and five nature managers (managing the natural park or the SAC). We tried to include stakeholders from the nine municipalities of the area (Figure 5), 16 in Galicia and 13 in Castile and Leon. In eight cases they do not live therein but nearby or they have political/administrative roles with a direct impact on the area. In order to guarantee anonymity, real names have been omitted and substituted by invented names, occasionally changing the genders. Most of these interviews have already been interpreted, in a paper solely focused on tourism, by Valerià Paül et al. (2019).

A semistructured interviewing schedule was created following Ruiz Olabuénaga's (1999), Rob Kitchin and Nicholas J. Tate's (2000), and Gill Valentine's (2005) conventions. The words "mountain(s)" and "Trevinca" were deliberately avoided to encourage participants to freely refer to these concepts (Cameron 2016). Through 25–30 questions phrased as "Tell me about", four themes were embraced:

- perceptions of the region;
- identification of people, institutions, organizations, and such working in the region, to detect actors and networks;
- assessment of the role played by the public sector, particularly related to policies impacting and promoting the region; and
- presence of an administrative limit between Galicia and Castile and Leon, trying to clarify to what extent there might be a divergence between both sides of the border.

The interviews lasted approximately one hour and, with permission, were recorded and transcribed. This material has been analyzed by means of open coding, both emic and etic (Crang 2005; Cope 2016). These codes are the basis for the results section of this paper, which are structured into three narratives obtained from semiotic clustering, carried out according to Mike Crang (2005). The quotes are our translation.

Results

The narratives are an interpretation of a consistent set of analytical codes regarding the Trevinca massif. Interviewees are not grouped by narratives because one interviewee may refer to codes attributed to different narratives. From a spatial perspective, we have inferred that each narrative has taken shape on particular hiking trails, embracing specific spatial ways of conceiving, planning, and managing them (Figure 7).

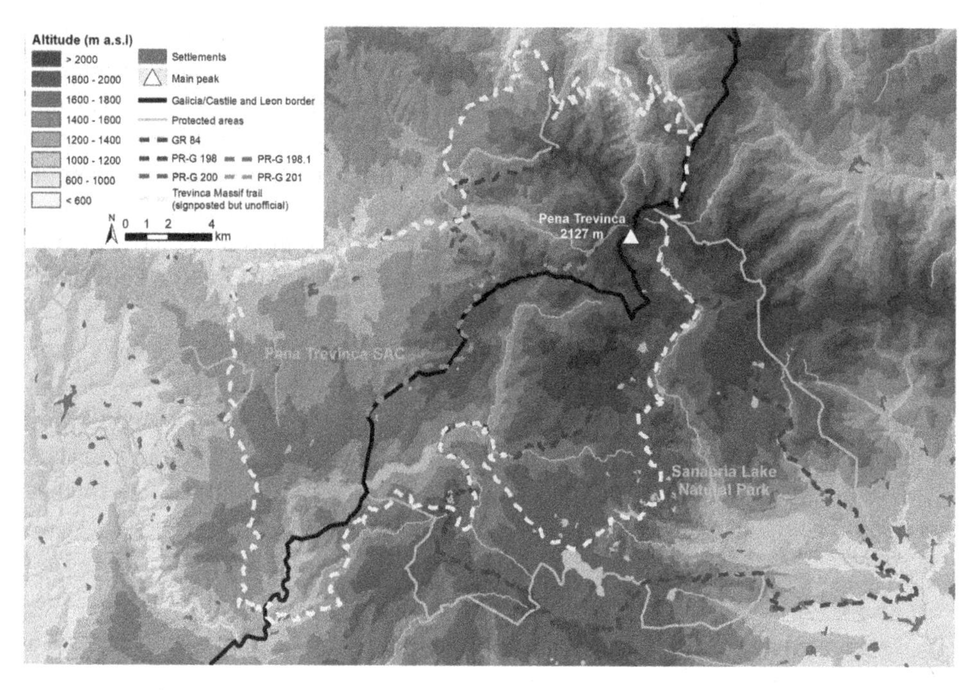

FIG. 7—Hiking trails associated with each narrative. Source: Base map from https://www.ign.es/, authors' elaboration.

WILDERNESS HIGHLANDS TO BE MOUNTAINEERED AND DESERVING INTEGRAL PROTECTION

The first narrative is based on the understanding of these mountains as areas of wilderness, with limited human intervention.

> It is an isolated area, without roads, [...] and, above all, undoubtedly for me, still pristine. [...] I walk every day [...] and for me it is like seeing a garden [Elvia].

> I would define the area as the last natural paradise [...]. Because there are some areas that are practically inaccessible, that still hold the essence of the mountain nature [Xoel].

These mountains are also seen as the perfect setting for achieving solitude and personal introspection.

> The highlands, mountainous, are spectacular. They amaze me. [...] The other day I was in [an upper valley], I stopped walking ... and I couldn't hear anything at all: simply the sound of nature [...]. A place to get lost ... or to find yourself [André].

> I often hike in the mountains. [...] It is relaxing, it helps me. [...] It is a therapy that I recommend [Iago].

Accordingly, the pivotal way to delve into the region is walking. In this sense, trails are relevant to access what is considered the soul of the region: the Trevinca peak. A revealing initiative, signposted in the early 2000s but not officially approved as a GR,[5] was the Trevinca Massif trail (Figure 7), a 135 km walk inspired by the well-known *Tour du Mont Blanc Randonnée.*

> Knowing the idea of the Mont Blanc Massif, [...] it could [...] interrelate all the places in the region [...] always in the lowlands [of the massif] [...]. This trail, afterwards, should have a kind of access to the highest parts, to the summits. [...] Three to four small huts were planned, open, for hikers [...]. [In some parts] the trail allows the view of Trevinca on the horizon, fantastic, [...] you see Trevinca covered in snow [Filipe].

Trevinca as a peak and the mountainous nature of the region are seen as central both to regional identity and at an international level (Figure 8). "Trevinca is the highest summit in Galicia, [...] a distinguishing mark for the region" [Brandán]; "[Trevinca] is the meeting point. It is like the symbol that has all the meaning there inside for us" [Tomé]. In this respect, the notion of "Galician Highlands" as a brand was conceived in this instance, a label that was subsequently appropriated by a particular local government, as will be explained later.

The extension of these highlands knows no borders. The massif is seen as a whole. "I don't believe in borders. I don't believe that beyond Trevinca [...] something different begins [...]. There is continuity between both sides [...] [T]he region [...] is the same" [Tomé].

Hence, a natural park for the whole region is upheld. Taking into account that there is no cross-border cooperation in place and that the Galician side has

FIG. 8—A group of mountaineers having climbed the Trevinca peak, unfolding a poster saying "For environmental and social justice. World Environment Day" (our translation). Source: Picture by Valerià Paül (4/6/2016).

not been designated, it seems essential that the massif as a whole attains common protection. "I would like this to be designated an interregional natural park. [...] We have a mountain that is shared, forests that are shared, rivers, valleys, lakes ... it is the same!" [Catuxa].

TREVINCA MOUNTAINS AS A TOURISM RESOURCE APPROPRIATED BY A VEIGA'S LOCAL GOVERNMENT

It is highly revealing that some interviewees, when mentioning "Trevinca," do not refer to the mountain(s), but to the unbuilt ski resort mentioned earlier.

> Trevinca was the most important project [discussed in the early 2000s] [...], but in the end, nothing came of it. They wanted to build a ski resort or something like that [Roi].

> Then we had a project for Trevinca, but there was nowhere to accommodate visitors, and with this story about Trevinca [...] [it was possible] to develop tourism [Olalla].

This is evidence of a vision conceiving the massif as a potential tourism attraction ready to be intensively developed. Improving accessibility is a milestone of this narrative.

> The ski resort was a blow to the whole area. [...] It was just an initial rallying point, and from there ... [...] I think the only potential for this area is to first have good road infrastructure [...]. And attract tourists [Sabela].

We identify two spatial scales attached to this conception. The first is older, from the early 2000s, and consisted of a LEADER project managed by several local governments belonging to both autonomous regions (Figure 9). Fundamentally devoted to tourism, the project initially focused on the ski resort. When this did not happen, it diversified amongst several tourism accommodation investments. "With that project we decided to develop hotels and rural tourism businesses to attract more tourists" [Olalla].

Given that this LEADER device was led by A Veiga's local government, this municipality has gained momentum, increasingly attaching its toponym to Trevinca. Although the massif is shared between different municipalities, A Veiga is making more profit from Trevinca, understood as a resource, with the declared object of converting it into an economic product. A Veiga's local government used "Galician Highlands," an expression originally from mountaineering (see above), to promote the municipality in the 2000s and, throughout the 2010s, it has consistently marketed the area. A case in point is the achievement of the "Starlight Destination" label[6] for the municipality of A Veiga with the name "Trevinca-A Veiga" (Figure 10). This designation has attracted widespread attention.

> The only one [local government] who made it [in the tourism industry, within the region at large] is A Veiga with its Starlight certification. [...] I think it [the Starlight campaign] will

FIG. 9—Signpost in Viana do Bolo showing the region covered by the LEADER intervention area. Source: Picture by Valerià Paül (26/1/2019).

> be very beneficial; it will make the area known, whether directly or indirectly it is going to help the area [...]. It's a form of advertising [Brais].

In 2016, A Veiga's local government funded four GRs (PR-G 198, PR-G 198.1, PR-G 200, PR-G 201). They reach the summit of Trevinca from different hamlets of the municipality, and are the first signs that have been blazed in the highest part of the mountains (Figure 7). Beyond tourism, these developments in A Veiga embrace other sectors by means of converting Trevinca into a trademark.

> The marketing and promotion of an area encompasses everything. Today a liter of honey is worth more [because it is labeled 'Trevinca'] [...]. Once we have created the Trevinca brand, all tourism, livestock, honey ... products with this trademark will increase in value [Hadrián].

These advances, led by A Veiga's local government, are discursively placed under the sustainability approach. However, some interviewees point out that they are

FIG. 10—The welcome signpost when entering the town of A Veiga says Trevinca-A Veiga rather than A Veiga. Source: Picture by Valerià Paül (5/11/2016).

incompatible with increasing levels of nature protection. "[The designation of the Natural Park] might be led by [A Veiga's] Local Government, but they are not willing to face hunters" [Xil].

THE SANABRIA LAKE NATURAL PARK AS A TOURISM DESTINATION, WITH TREVINCA AS A DISTANT PEAK CONSTITUTING THE BORDER WITH GALICIA

This narrative refers to the same landscape, but from the other side, instead of from a Galician perspective. The massif is seen as a remote and inaccessible area.

> Getting there is very difficult. [...] Trevinca is the distant aim for many people [...]. It is the highest peak, the place which is furthest away. It requires an effort. It is an emblematic place for a lot of people, as it is challenging, and not a lot of people go there [Martiño].

These mountains are perceived as tough, as an endeavor because of the hard environmental conditions.

> [To get to Trevinca] is the longest trail in terms of time that we have here [...], eight to ten hours. [...] You must be very careful [Xonxa].

> A footbridge has fallen in the mountains [...] and in the mountains this [the infrastructure] is constantly damaged [Millao].

This perspective of mountains being far away is derived from an understanding of the Sanabria Lake as the center of the region. From a tourism perspective, the lake might be connected with the mountains in the future.

> We should encourage visitors to leave the lake and go to other places. [...] [The mountains] have a large capacity to receive visitors [Martiño].

> The government of Castile and Leon is discussing installing a funicular [connecting the lake level, 1050 m, with the top of a neighboring hill, 1600 m] [...]. This would give the area a boost, it would become an important attraction, revitalizing the hamlets [Nuno].

The notion that these mountains are foreseen as part of an expanded tourism destination is consistent with the general agreement that the lake itself has been over-touristified, experiencing what is labeled as "mass tourism" [Filipe]. "There are more and more bars, hostels, hotels ... and everything is competition. There are activities that there didn't use to be [...]. It's too much" [Xonxa].

The park has marked a GR (GR 84) structuring the protected area without connections with the environs (Figures 7 and Figure 11). Accordingly, the park does not interact with Galicia. "Relations [with Galicia] are ... scant. [...] We should have more cooperation [...] as it is the same area, it is exactly the same" [Martiño]. Trevinca becomes, then, a dividing, and even invisible, mountain. "[Trevinca] is the peak that divides us" [Paulo]. "There are people living here [...] who have never been to [...] Trevinca nor do they even know that it exists" [Tomé].

Discussion and Conclusions

This paper analyzes the discursive emergence of the Galician Highlands as a spatial imaginary, trying to decipher whether this leads toward a new Galician national landscape of its own and/or whether this responds to global forces. Three narratives have emerged from interviewing, contributing unequally to this construction, as will be further argued. Each one of them seems consistent with either one or several of the Western imaginaries regarding mountains, as outlined by Debarbieux (2004) and Debarbieux and Rudaz (2010), amongst others.

Firstly, the narrative understanding the massif as a place of wilderness to be protected because of its outstanding natural and spiritual values is in accordance with the enlightened and romantic imaginaries of the mountains—that is, the era of the Western "invention of the mountain" (Joutard 1986; Short 1991; Martínez de Pisón and Álvaro 2002; Sacareau 2003; Debarbieux 2004; Debarbieux and Rudaz 2010). The association with mountaineering is highly relevant, not only in sportive terms—such as hiking—but, more transcendentally, in the sense of establishing emotional connections (Martínez de Pisón and Álvaro 2002; Nogué and Vicente 2004; Bernbaum and Price 2013). There is also a link with the wilderness vision (Short 1991; Cronon 1995; Depraz 2008; Debarbieux and Rudaz 2010; Woods 2011;

Fig. 11—A GR 84 blazed signpost, with the Trevinca peak, partially covered with clouds, in the background. The closest place that GR 84 is to the peak is 5 km (Figure 7). Source: Picture by Valerià Paül (3/6/2017).

Debarbieux 2019). A further connection relates to the global mountain imaginary (Debarbieux and Rudaz 2010; Price and Kohler 2013), as evidenced in Figure 8.

This first narrative might have connections with nationalist arguments in accordance with theorizations by authors such as Nogué and Vicente (2004), Debarbieux and Rudaz (2010), Herb (2018), and Paül (2019). However, we only find a feeble vindication in this regard from the mountaineering precedents of the 1940s–50s, developed in tough political circumstances that complicated the creation and dissemination of cultural products (Figure 12), as found in other cases analyzed by Martínez de Pisón and Álvaro (2002), Debarbieux and Rudaz (2010), and Della Dora (2016). Moreover, nowadays, there is only a very moderate use of Galician national symbols—such as the flag—in current mountaineering activities. A possible underlying factor is that, contrary to other countries (Nogué and Vicente 2004; Debarbieux and Rudaz 2010; Herb 2018), Galicia lacks its own landscape tradition linked to mountains from a national perspective.

The "Galician Highlands" notion was conceived in this first instance. Interestingly, this notion was named to resonate with Galician identity concerns—in an attempt to enable the massif to be assumed at the national level. Ironically, the notion can only be understood within some immanent global connections, as the acknowledged inspiration is the Scottish region toponym,

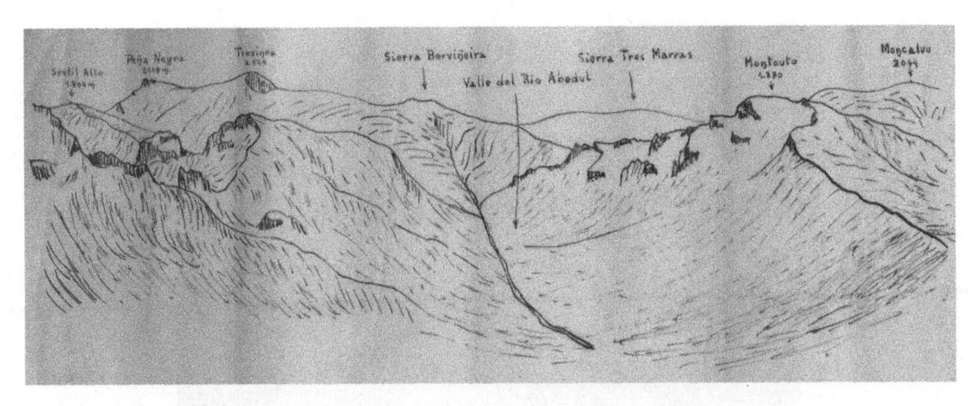

FIG. 12—Trevinca in the mid-1940s, as seen by the mountaineer Francisco Marfany, Gonzalo Gurriarán's friend. This is an unpublished drawing taken from the mostly unknown literature on the Galician highlands which was developed in the 1940s-50s. Source: Ricardo Gurriarán's personal archive (reproduced with permission).

confirming that the Scottish Highlands landscape construction has had a powerful capacity to frame a spiritual notion of distinctiveness at a world scale (Perkins 2006; Debarbieux and Rudaz 2010). In any case, the spread of this spatial category has been developed in the context of the second narrative, thus becoming a social imaginary, as the first narrative has had a humble capacity to influence wider imaginaries.

Secondly, the narrative conceiving the Trevinca Massif as a potential resource appropriated by political entities, especially for tourism purposes, matches the modernist imaginary of mountains as policy-making regions for developing economic activities (Majoral 1997; Pujadas and Font 1998; Debarbieux 2004; Debarbieux and Rudaz 2010). In this context, the Trevinca 2000–2006 LEADER device can be read as an institutionalization of mountain regions, described by Debarbieux and Rudaz (2010) and Rob McMorran et al. (2015). However, in the 2010s, this massif-scale LEADER device vanished and the narrative has since been monopolized by A Veiga's local government, which has equalized the municipality with the massif as a whole.

Although we have found attempts to mobilize the primary sector based on the mountain notion, tourism is central to this second narrative. Beyond Paül et al. (2019) examination of the emergence of Trevinca–A Veiga as a tourism destination, this research bears witness to the ability of tourism to determine a spatial imaginary, commodifying a rural landscape (Tonts and Greive 2002; Perkins 2006; Woods 2011). By means of marketing, labeling development, and meetings of popular influencers on social networks, A Veiga's local government has been able to position Trevinca and the Galician Highlands at a Galicia-wide level and even to promote the mountains as a landmark for Galicia abroad (Figure 13). Additionally, with the Starlight brand, global marketing has been

FIG. 13—#Trevinca on the official tourism Twitter account of the Galician Government: a glacial lake and the superbloom, with the hashtags #GaliciaYouHaveToLiveIt #ChooseGalicia. Source: @turgalicia (31/7/2018) (reproduced with permission from Turismo de Galicia).

reinforced and now the area is proudly part of a worldwide network of certified regions without light pollution.[7] This is a worthy addition to the literature on mountain tourism (see Nepal and Chipeniuk 2005; Price and Kohler 2013). Hence, tourism should not only be understood as an economic activity in the mountains, but also as a relevant procedure to create and impel their spatial imaginaries, implying a global dimension.

Thirdly, the Sanabria Lake-centered narrative responds to the inception of protection devices in mountain areas (Depraz 2008; Debarbieux and Rudaz 2010; Price and Kohler 2013; Arpin and Cosson 2015). This confirms the capacity of protected areas, especially national and natural parks, to create discourses that can become spatial imaginaries. The development of this narrative clearly places tourism in its core but spatially overfocusing on the lake, with limited capacity to incorporate the massif. Although we acknowledge a will to access Trevinca and the highest areas, the narrative itself favors a vision of the mountains as remote, difficult, and so forth.

There is an inherent tension between the first narrative and the third one, as the former demands more protection by only allowing harmless forms of hiking, while the latter envisages infrastructure and investments linked to increasing levels of tourism. This confirms the widespread contest between tourism and protection, resulting in multiple solutions (Nepal and Chipeniuk 2005; Depraz 2008; Price and Kohler 2013; Arpin and Cosson 2015).

This third narrative can also be interpreted under the lens of the interaction between borders and mountains. Indeed, this narrative does not participate in the social imaginary of the Galician Highlands, as it is detached from Galicia. In other words, although the Galician nationalist construction is feeble, as explained above, it has ultimately contributed to the spatial divorce dividing the massif into two parts: the western part has become, in a sense, a "nationalized landscape," and the eastern part is basically a natural park. Past case studies report the disputed notion of "natural borders" coinciding with mountains (Sahlins 1990; Debarbieux and Rudaz 2010; Fall 2010), but this research makes it evident that this is an immanent, ongoing debate. In fact, the contrast between [Paulo] saying Trevinca divides, while [Tomé] believing it unites is highly revealing.

The separation between the emergent imaginary of the Galician Highlands within Galicia, on the one hand, and the Sanabria Lake Natural Park, on the other, bears witness to the lack of cooperation between neighboring, devolved, autonomous governments within Spain, as already described by Joan Romero (2009, 2012). In fact, the first narrative analyzed here suggests the idea of global protection for the massif as a whole. Debarbieux and Rudaz (2010), Jörg Balsiger and Stéphane Nahrath (2015), Charles C. Chester (2015), and Debarbieux et al. (2015) have reported cases where mountains have been the instance for cooperation. Why this is possible between neighboring countries but not in our case-study area, where the border corresponds to an internal limit between two devolved territorial entities within Spain, remains an open question.

The overall interpretation of the obtained narratives does not dispute the model of Western succession of mountain imaginaries by Debarbieux (2004) and Debarbieux and Rudaz (2010). However, we stress that this background literature does not take into consideration the narratives obtained from interviewing, reducing the ability to capture nuances and contradictions on the ground, as expected by Ruiz Olabuénaga (1999) and Valentine (2005). In fact, the dominant literature on mountain imaginaries seems largely based on elitist cultural products. However, "to prevent the social imaginary from being a kind of floating abstraction with a vague status, it has to be attached to concrete practices, including those of individual imagination" (Debarbieux 2019, 4). We therefore conclude that in order to grasp social imaginaries regarding the mountains and their national and/or global ascriptions, it is essential to start from the views of individuals. Importantly, we also infer that imaginations can have a material consequence, as expressed in Figure 7, where every hiking trail is the shape of a particular narrative.

This paper has shown that the emergence of the Galician Highlands as a new socio-spatial imaginary about Galicia responds to two interrelated processes. On the one hand, the overall preference for coastal and farming landscapes in Galicia is being nuanced with the appearance of the mountainous landscape that we have analyzed here. Hence, the Galician national landscape imaginaries are experiencing a reconfiguration in relation to the inherited visions discussed above. On the other hand, the rise of the Galician Highlands construct has been impelled by global forces, namely tourism and branding for promotion purposes. In this regard, the ultimate construction of a new national landscape for Galicia based on mountains, although using some national attributes (for example, the name of Galicia itself), seems more related to the need of positioning Galicia internationally, embracing an inherent global dimension. In the end, it can be asserted that Galicia has been incorporated into the club of many Western countries with a mountain imaginary. Thus, the recent materialization of a deliberate artistic project on Galician mountains, led by photographer Fernández Pérez de Lis (2017), evidences that the mountains—with Trevinca at the forefront—have definitively become a Galician spatial imaginary.

Acknowledgments

We acknowledge all the interviewees for their time and their insightful views. We thank Prof Holly Barcus and Prof William Moseley, who led the special issue where this paper is published, after the 27th IGU CSRS Annual Colloquium held in Minnesota and Wisconsin (USA) in July 2019, including their review of an early draft of this manuscript. We are grateful for the helpful comments and criticisms made by the three anonymous referees. The authors would like to thank Carmel Sherrington and Prof Laura Lojo for their work editing the English on different versions of this paper. We also acknowledge Alejandro Gómez Pazo for his support producing the maps through intensive GIS effort and Dr Ricardo Gurriarán for granting us access to his personal archive. The customary disclaimers apply.

Funding

The fieldwork necessary to carry out these interviews was funded by the Galician Government (Project n. 2016-PG009: "Territorial Cooperation in the Galician Borderlands: Analyzing External and Internal Cross-border Governance").

Notes

1. Following Debarbieux (2004), the attention here concentrates on Western countries, although the West is a very problematic concept (Lewis and Wigen 1997).
2. Under LEADER, specific rural regions are selected to encourage development projects and actions based on local strategies.
3. Except in the case of Porto, the other municipalities have more than one hamlet in their local government area. In order to refer to the area as a whole, we favor the municipal scale (with nine municipalities shown in Figures 3 and 5).
4. An EU network of protected areas. Despite the alleged consistency across the EU, protection is a national (in Spain, regional) responsibility. See https://ec.europa.eu/environment/nature/natura2000/index_en.htm
5. GR is the acronym for *Grande Randonnée* in French, a European-wide system of official long-distance footpaths blazed with characteristic marks consisting of white and red stripes.
6. It is awarded by the Starlight Foundation (Institute of Astrophysics of the Canary Islands).
7. https://www.fundacionstarlight.org/en/section/list-of-starlight-tourist-destinations/293.html.

Orcid

Valerià Paül http://orcid.org/0000-0003-3007-1523

Juan-M. Trillo-Santamaría http://orcid.org/0000-0003-3842-3079

References

Arpin, I., and A. Cosson. 2015. The Category of Mountain as Source of Legitimacy for National Parks. *Environmental Science & Policy* 49:57–65. doi: 10.1016/j.envsci.2014.07.005.

Balsiger, J., and S. Nahrath. 2015. Functional Regulatory Spaces and Policy Diffusion in Europe: The Case of Mountains. *Environmental Science & Policy* 49:8–20. doi: 10.1016/j.envsci.2015.01.004.

Beramendi, J. 2007. *De provincia a nación. Historia do galeguismo político*. Vigo, Spain: Xerais.

Bernbaum, E., and L. W. Price. 2013. Attitudes Towards Mountains. In *Mountain Geography: Physical and Human Dimensions*, edited by M. F. Price, et al., 253–266. Berkeley: University of California Press.

Byers, A. C., L. W. Price, and M. F. Price. 2013. An Introduction to Mountains. In *Mountain Geography: Physical and Human Dimensions*, edited by M. F. Price, et al., 1–10. Berkeley: University of California Press.

Cameron, J. 2016. Focusing on the Focus Group. In *Qualitative Research Methods in Human Geography*, edited by I. Hay, 203–224. Don Mills, Canada: Oxford University Press.

Chester, C. C. 2015. Yellowstone to Yukon: Transborder Conservation across a Vast International Landscape. *Environmental Science & Policy* 49:75–84. doi: 10.1016/j.envsci.2014.08.009.

Cope, M. 2016. Organizing and Analyzing Qualitative Data. In *Qualitative Research Methods in Human Geography*, edited by I. Hay, 373–392. Don Mills, Canada: Oxford University Press.

Crang, M. 2005. Analysing Qualitative Materials. In *Methods in Human Geography*, edited by R. Flowerdew and D. Martin, 218–232. Harlow, UK: Prentice Hall.

Cronon, W. 1995. The Trouble with Wilderness; Or, Getting Back to the Wrong Nature. In *Uncommon Ground: Rethinking the Human Place in Nature*, 69–90. New York: W. W. Norton & Co.

Cunha, S. F., and L. W. Price. 2013. Agricultural Settlement and Land Use in Mountains. In *Mountain Geography: Physical and Human Dimensions*, edited by M. F. Price, et al., 301–331. Berkeley: University of California Press.

De Castro, R. 1863. *Cantares Gallegos*. Vigo, Spain: Juan Compañel.

Debarbieux, B. 2004. The Symbolic Order of Objects and the Frame of Geographical Action: An Analysis of the Modes and Effects of Categorisation of the Geographical World as Applied to the Mountains in the West. *GeoJournal* 60 (4):397–405. doi: 10.1023/B:GEJO.0000042976.00775.24.

Debarbieux, B., and G. Rudaz. 2010. *Les faiseurs de montagne: Imaginaires politiques et territorialités XVIIIe-XXIe siècle*. Paris: Centre National de la Recherche Scientifique.

Debarbieux, B., M. F. Price, and J. Balsiger. 2015. The Institutionalization of Mountain Regions in Europe. *Regional Studies* 49 (7):1193–1207. doi: 10.1080/00343404.2013.812784.

———. 2019. *Social Imaginaries of Space. Concepts and Cases*. Cheltenham/Northampton, UK: Edward Elgar.

Della Dora, V. 2016. *Mountain: Nature and Culture*. London: Reaktion Books.

Depraz, S. 2008. *Géographie des espaces naturels protégés. Genèse, principes et enjeux territoriaux*. Paris: Armand Colin.

Fall, J. J. 2010. Artificial States? On the Enduring Geographical Myth of Natural Borders. *Political Geography* 29 (3):140–147. doi: 10.1016/j.polgeo.2010.02.007.

Fernández Pérez de Lis, J. 2017. *Catro ventos=Four Winds*. Milladoiro, Spain: Fabulatorio.

Gurriarán, R. 2005. *Gonzalo Gurriarán Gurriarán (1904–1975)*. O Barco de Valdeorras, Spain: Peymar.

Herb, G. H. 2018. Power, Territory, and National Identity. In *Scaling Identities: Nationalism and Territoriality*, edited by G. H. Herb and D. H. Kaplan, 7–30. Lanham, UK: Rowman & Littlefield.

Joutard, P. 1986. *L'invention du Mont Blanc*. Paris: Gallimard/Julliard.

Kitchin, R., and N. J. Tate. 2000. *Conducting Research into Human Geography. Theory, Methodology and Practice*. New York: Routledge.
Lewis, M. W., and K. Wigen. 1997. *The Myth of Continents: A Critique of Metageography*. Berkeley: University of California Press.
Lois, R. C. et al. 2015. The Way of Saint James: A Contemporary Geographical Analysis. In *The Changing World Religion Map*, edited by S. Brunn, 709–732. Dordrecht, NL: Springer.
López Sández, M. 2008. *Paisaxe e nación. A creación discursiva do territorio*. Vigo, Spain: Galaxia.
López Silvestre, F. 2004. *El discurso del paisaje. Historia cultural de una idea estética en Galicia (1723–1931)*. Santiago de Compostela, Spain: Universidade de Santiago de Compostela.
Majoral, R. 1997. Desarrollo rural en zonas de montaña. *Geographicalia* 34:23–49.
Martínez de Pisón, E., and S. Álvaro. 2002. *El sentimiento de la montaña. Doscientos años de soledad*. Madrid: Desnivel.
Mato, M. 1998. *A escrita da terra. Configuración do espacio natural na literatura galega*. A Coruña, Spain: Espiral Maior.
McMorran, R., et al. 2015. A Mountain Food Label for Europe?. *Journal of Alpine Research* 103 (4). doi: 10.4000/rga.2654.
Miguélez-Carballeira, H. 2014. *Galiza, um povo sentimental? Género, política e cultura no imaginário nacional galego*. Santiago de Compostela, Spain: Associaçom Galega da Língua.
Murado, M. A. 2008. *Otra idea de Galicia*. Barcelona: Debate.
Nepal, S. K., and R. Chipeniuk. 2005. Mountain Tourism: Toward a Conceptual Framework. *Tourism Geographies* 7 (3):313–333. doi: 10.1080/14616680500164849.
Nogué, J., and J. Vicente. 2004. Landscape and National Identity in Catalonia. *Political Geography* 23 (2):113–132. doi: 10.1016/j.polgeo.2003.09.005.
Paül, V. 2019. Catro breves hipóteses na interface entre paisaxe e nación en Galicia. In *Paisaxes nacionais no mundo global*, edited by J.-M. Trillo-Santamaría and R. C. Lois, 83–109. Santiago de Compostela, Spain: Grupo de Análise Territorial.
Paül, V., J.-M. Trillo-Santamaría, and F. Haslam Mckenzie. 2019. The Invention of a Mountain Tourism Destination: An Exploration of Trevinca-A Veiga (Galicia, Spain). *Tourist Studies* 19 (3):313–335. doi: 10.1177/1468797619833364.
Perkins, H. K. 2006. Commodification: Re-Resourcing Rural Areas. In *Handbook of Rural Studies*, edited by P. Cloke, T. Marsden, and P. Mooney, 243–257. Thousand Oaks, CA: SAGE.
Price, M. F., and T. Kohler. 2013. Sustainable Mountain Development. In *Mountain Geography: Physical and Human Dimensions*, edited by M. F. Price, et al., 333–365. Berkeley: University of California Press.
Pujadas, R., and J. Font. 1998. La ordenación de las áreas de montaña. In *Ordenación y planificación territorial*, 265–283. Madrid: Síntesis.
Romero, J. 2009. *Geopolítica y gobierno del territorio en España*. València, Spain: Tirant lo Blanch.
———. 2012. España inacabada: Organización territorial del Estado, autonomía política y reconocimiento de la diversidad nacional. *Documents d'Anàlisi Geogràfica* 58 (1):13–59. doi: 10.5565/rev/dag.190.
Ruiz Olabuénaga, J. I. 1999. *Metodología de la investigación cualitativa*. Bilbao, Spain: Universidad de Deusto.
Sacareau, I. 2003. *La montagne: Une approche géographique*. Paris: Belin.
Sahlins, P. 1990. Natural Frontiers Revisited: France's Boundaries since the Seventeenth Century. *The American Historical Review* 95 (5):1423–1451. doi: 10.2307/2162692.
Santos, X. M., and J.-M. Trillo-Santamaría. 2017. Tourism and Nation in Galicia (Spain). *Tourism Management Perspectives* 22:98–108. doi: 10.1016/j.tmp.2017.03.006.
Schulz, G. 1835. *Descripción geognóstica del Reino de Galicia, acompañada de un mapa petrogr[á]fico de este país*. Madrid: Imprenta de los Herederos de Collado.
Short, J. R. 1991. *Imagined Country: Environment, Culture and Society*. Syracuse, NY: Syracuse University Press.
Tonts, M., and S. Greive. 2002. Commodification and Creative Destruction in the Australian Rural Landscape: The Case of Bridgetown, Western Australia. *Australian Geographical Studies* 40 (1):58–70. doi: 10.1111/1467-8470.00161.
Valentine, G. 2005. Tell Me about ... : Using Interviews as a Research Methodology. In *Methods in Human Geography*, edited by R. Flowerdew and D. Martin, 110–127. Harlow, UK: Prentice Hall.
Woods, M. 2011. *Rural*. New York: Routledge.

Index

Note: Figures are indicated by *italics*. Tables are indicated by **bold**. Endnotes are indicated by the page number followed by 'n' and the endnote number e.g., 20n1 refers to endnote 1 on page 20.